Tactical Display for Soldiers
Human Factors Considerations

Panel on Human Factors in the Design of Tactical Display Systems
for the Individual Soldier

Committee on Human Factors

Commission on Behavioral and Social Sciences and Education

National Research Council

NATIONAL ACADEMY PRESS
Washington, D.C. 1997

NATIONAL ACADEMY PRESS • 2101 Constitution Avenue, NW • Washington, DC 20418

NOTICE: The project that is the subject of this report was approved by the Governing Board of the National Research Council, whose members are drawn from the councils of the National Academy of Sciences, the National Academy of Engineering, and the Institute of Medicine. The members of the committee responsible for the report were chosen for their special competences and with regard for appropriate balance.

This report has been reviewed by a group other than the authors according to procedures approved by a Report Review Committee consisting of members of the National Academy of Sciences, the National Academy of Engineering, and the Institute of Medicine.

The work of the Panel on Human Factors in the Design of Tactical Display Systems for the Individual Soldier is supported by Department of the Army Contract No. DAAD05-92-C-0087 issued by the U.S. Army Research Laboratory. The views, opinions, and findings contained in this report are those of the authors(s) and should not be construed as an official Department of the Army position, policy, or decision, unless so designated by other official documentation.

Library of Congress Cataloging-in-Publication Data

National Research Council (U.S.). Panel on Human Factors in the
Design of Tactical Display Systems for the Individual Soldier.
Tactical display for Soldiers : human factors considerations /
Panel on Human Factors in the Design of Tactical Display Systems for
the Individual Solider, Committee on Human Factors, Commission on
Behavioral and Social Sciences and Education, National Research
Council.
p. cm.
Includes bibliographical references and index.
ISBN 0-309-05638-1
1. Helmet-mounted displays. 2. United States. Army. Infantry—
Equipment. 3. Human engineering. I. Title.
UG489.N38 1997
623.4′6—dc21 96-39712
CIP

Additional copies of this report are available from National Academy Press, 2101 Constitution Avenue NW, Box 285, Washington, DC 20055. 800-624-6242 or 202-334-3313 in the Washington metropolitan area. http://www.nap.edu.

Printed in the United States of America

PANEL ON HUMAN FACTORS IN THE DESIGN OF TACTICAL DISPLAY SYSTEMS FOR THE INDIVIDUAL SOLDIER

WILLIAM O. BLACKWOOD (*Chair*), Hay Management Consultants, Inc., Arlington, Virginia
TIMOTHY R. ANDERSON, Armstrong Laboratory, Wright-Patterson Air Force Base, Ohio
C. THOMAS BENNETT, Lawrence Livermore Laboratory, University of California, Berkeley
JOHN R. CORSON, Consultant, Williamsburg, Virginia
MICA R. ENDSLEY, Department of Industrial Engineering, Texas Tech University, Lubbock
PETER A. HANCOCK, Human Factors Research Laboratory, University of Minnesota
JULIAN HOCHBERG (emeritus), Department of Psychology, Columbia University
JAMES E. HOFFMAN, Department of Psychology, University of Delaware
RONALD V. KRUK, CAE Electronics, Ltd., Saint Laurent, Quebec, Canada

ANNE S. MAVOR, Study Director
JERRY S. KIDD, Senior Staff Officer
CINDY S. PRINCE, Senior Project Assistant

COMMITTEE ON HUMAN FACTORS

Contents

Preface

The Panel on Human Factors in the Design of Tactical Displays for the Individual Soldier was established by the National Research Council at the request of the U.S. Army Natick Research, Development, and Engineering Center for the purpose of explicating the human factors issues and approaches associated with the development, testing, and implementation of helmet-mounted display technology in the Land Warrior System. More specifically, the panel was charged with examining the relationship among the tactical information needs of individual soldiers; the possible devices available now and in the near future for processing, transmitting, and displaying such information; and the human performance implications of the use of such devices.

This report presents our analysis, findings, and recommendations. The requirement for the proposed Land Warrior System stems from the need to help the soldier think and act quickly and effectively. Underlying the design concept is the assumption that the resulting system will improve the soldier's situation awareness of the battlefield and by so doing increase his performance efficiency and accuracy. In order to appropriately frame the critical human factors issues regarding design, training, and conditions of use, we begin by characterizing the sources of the requirement—the mission needs and the infantry soldier. Chapter 1 discusses military context, infantry doctrine, and employment scenarios for the proposed Land Warrior System; Chapter 2 presents the characteristics of the intended user. Designing usable equipment and providing effective training requires a detailed knowledge of the potential uses and the users as well as the possible interactions between design attributes and user capabilities and limitations. Because situation awareness is a central goal of the system, we begin our technical analysis in Chapter 3 by considering how the land warrior helmet-mounted display may have both positive and negative impacts on the soldier's awareness of his surroundings and in turn on his ability to act. Effective situation awareness of the battlefield requires not only knowledge of friendly and enemy forces but also understanding of the dynamics of actions expected to occur in the near future. Since the soldier's situation awareness and performance are directly affected by how well he sees and interprets images of the world, it is essential to consider the proposed display and its implications for viewing and moving in the environment.

The eyes are the primary input channel for the soldier. When using the proposed system at night, the soldier will view the battlefield, detect and engage targets, and perform all his soldiering tasks through the display. Under both nighttime and daytime conditions, the soldier will read, respond to, create, and interpret images and messages that are transmitted over the system. Chapter 4 addresses the visual factors in designing and evaluating display devices in terms of the visually guided behaviors and tasks they are intended to support.

Like the eyes, the ears are essential to the soldier's performance and survivability. Auditory communications on the battlefield have always been challenging because of interference from background noise, the imprecision of oral communication, and fatigue. Our discussion of auditory displays in Chapter 5 focuses on supplementing or amplifying visual information, evaluating the pros and cons of various types of auditory signals, and examining the conditions and tasks for which auditory information is most effective.

Our final area of technical analysis, covered in Chapter 6, is the potential effects of the proposed Land Warrior helmet-mounted display on the soldier's cognitive workload. The Land Warrior System will add new tasks for the soldier to perform that will require him to acquire more information through reading and to process that information for retransmission, decision making, or both. It will be important for the Army to consider the implications of this additional workload on the performance of infantry soldiers who meet minimum recruiting standards.

As a result of our analysis of relevant research, the panel identified many areas in which additional knowledge is needed both for research and at the operational test and evaluation level. As a result, in Chapter 7 we identify a critical set of human factors test issues and propose a strategy for collecting information about these issues. This plan is followed in Chapter 8 by the conclusions and recommendations that the panel feels are most important. Some of these recommendations are in the form of design guidelines; others provide the basis for a proposed research agenda.

We begin our acknowledgments by extending thanks to Claire Gordon and Cynthia Blackwell, sponsor representatives from the U.S. Army Natick Research, Development, and Engineering Center (Natick RD&E) for their foresight in recognizing the importance of human factors for the Land Warrior System and for providing sponsorship and information support for the panel. Throughout the course of the project, many individuals have made contributions to the panel's thinking by serving as presenters and sources of information. We are particularly grateful to Bernard Corona, Human Research and Engineering Directorate, Army Research Laboratory, for his continued interest in the project and for providing us with his valuable insights on the Land Warrior System and the state of human factors in related technologies. We also received cooperation and support from numerous members of the Army staff at the Infantry School, Fort Benning, the Training and Doctrine Command, and the Army Materiel Command. We would like to extend our special thanks to Captain Gregory Dyekman of the Infantry School for taking care of our logistic needs and for scheduling special demonstrations of the Soldier Integrated Protective Ensemble (SIPE) and night vision equipment. Others who contributed during our visit to Fort Benning were: Major Marc Collins, Project Manager-Soldier Systems; Captain William Dickey, Major Ronald Murray, Sergeant Richardson, and Staff Sergeant Weiser of the Infantry School; Lieutenant Colonel Ross Holden, Captain Ed Jennings, and Captain Scott O'Neil of the Night Vision Laboratory; and Patrick Snow, Jr., U.S. Natick RD&E Center.

Although this report is the collective product of the entire panel, in the course of preparing it, each member of the panel took an active role in drafting sections of chapters, leading discussions, and reviewing successive drafts. In particular, John Corson provided expert knowledge on the military context and on planning for research, development, testing, and evaluation; Mica Endsley contributed extensively in the area of situation awareness; Julian Hochberg, James Hoffman, and Ronald Kruk assumed major responsibility for the work on visual and psychomotor issues; Timothy Anderson provided material on auditory displays; Peter Hancock contributed material on cognitive workload; and Tom Bennett provided expertise in decision making.

Staff at the National Research Council made important contributions to our work in many ways. We would like to express our gratitude to Jerry Kidd, senior staff officer, for his hard work and technical support; to Alexandra Wigdor, director of the Division on Education, Labor, and Human Performance, for her valuable guidance; and to Cindy Prince, the panel's administrative assistant, who was indispensable in organizing meetings, compiling agenda materials, and working on the final manuscript. We are also indebted to Christine McShane, who edited and significantly improved the report.

William O. Blackwood, Chair
Anne S. Mavor, Study Director

The National Academy of Sciences is a private, nonprofit, self-perpetuating society of distinguished scholars engaged in scientific and engineering research, dedicated to the furtherance of science and technology and to their use for the general welfare. Upon the authority of the charter granted to it by the Congress in 1863, the Academy has a mandate that requires it to advise the federal government on scientific and technical matters. Dr. Bruce M. Alberts is president of the National Academy of Sciences.

The National Academy of Engineering was established in 1964, under the charter of the National Academy of Sciences, as a parallel organization of outstanding engineers. It is autonomous in its administration and in the selection of its members, sharing with the National Academy of Sciences the responsibility for advising the federal government. The National Academy of Engineering also sponsors engineering programs aimed at meeting national needs, encourages education and research, and recognizes the superior achievements of engineers. Dr. William A. Wulf is interim president of the National Academy of Engineering.

The Institute of Medicine was established in 1970 by the National Academy of Sciences to secure the services of eminent members of appropriate professions in the examination of policy matters pertaining to the health of the public. The Institute acts under the responsibility given to the National Academy of Sciences by its congressional charter to be an adviser to the federal government and, upon its own initiative, to identify issues of medical care, research, and education. Dr. Kenneth I. Shine is president of the Institute of Medicine.

The National Research Council was organized by the National Academy of Sciences in 1916 to associate the broad community of science and technology with the Academy's purposes of furthering knowledge and advising the federal government. Functioning in accordance with general policies determined by the Academy, the Council has become the principal operating agency of both the National Academy of Sciences and the National Academy of Engineering in providing services to the government, the public, and the scientific and engineering communities. The Council is administered jointly by both Academies and the Institute of Medicine. Dr. Bruce M. Alberts and Dr. William A. Wulf are chairman and interim vice chairman, respectively, of the National Research Council.

Tactical Display for Soldiers

Human Factors Considerations

Executive Summary

THE LAND WARRIOR SYSTEM

The proposed helmet-mounted display of the Land Warrior System is one part of a major research and development effort by the U.S. Army to equip infantry soldiers for the high-technology battlefield of the future. The Mission Needs Statement for the 21st Century Land Warrior System (U.S. Department of the Army, 1993) calls for improvements in lethality, command and control, survivability, mobility, and sustainability in support of individual, dismounted infantry soldiers. The Land Warrior System is planned to provide vision enhancement (under both daytime and nighttime conditions), secure voice communication, greater protection, reduced load, and adequate support for individual maintenance in the tactical environment. The operational concept is stated as follows (U.S. Department of the Army, 1994):

> The [system] will be used by dismounted combat soldiers. The system will significantly enhance the soldier's ability to engage and defeat enemy targets while minimizing friendly casualties. The command, control, communications, computer and intelligence (C4I) subsystem will facilitate dynamic transmission of battlefield information, and enable soldiers to access the digitized battle field.

Guided by these requirements, military planners have conceptualized an ensemble of equipment that includes new protective garments, armaments, and information-processing elements. The visual display component of the information-processing subsystem is envisioned as a flip-down monocular presentation device mounted to one side of the soldier's helmet. The display proposed for daytime operations is an opaque display that can be used to provide the soldier with navigation information, such as maps and current location from the global positioning system; various command and control functions, such as messages regarding danger and troop movements; and images of the real world acquired through the thermal weapon sight. The display proposed for nighttime operations is integrated with a monocular night vision system or thermal weapon sight—when switched on, it appears in the middle of the night vision field of view. This display is used, as is the one designed for daytime, to provide digital data and weapon sighting information. These are two distinct pieces of equipment that are selected for use based on conditions.

ENHANCING THE PERFORMANCE OF INDIVIDUAL SOLDIERS

The Army's interest in providing individual, dismounted infantry soldiers with new tactical information systems has been partly motivated by combat experiences in recent conflicts. Specifically, the individual soldier's accomplishment of an assigned mission and his survival appear to be correlated with the amount and quality of information he is provided (Franks, 1994). The plans have also been influenced by advances in technology. Such advances include both hardware and software innovations in information-processing systems. Similarly, sensor technology has been changing. These changes are most apparent in the area of nighttime operations. Individual soldiers already have access to ambient light intensification and infrared detection equipment that has been used in combat, and those devices have shown sufficient promise to warrant continuation of their development.

At the same time, the entire domain of tactical intelligence has expanded. For example, satellites can now detect tactical targets and transmit the information almost instantaneously to command centers for relay to a combat unit. Other satellite arrays and packet switching[1] make it possible for soldiers to determine their location within a radius of a few meters from any site on the surface of the Earth with direct downlinks from satellites. When one's own location can be determined and that information is used in conjunction with laser range finding, it can be as if an individual soldier has whole batteries of artillery under direct control. Thus, overall, there appear to be opportunities to use the advances in technology to provide information that will help soldiers perform their missions, avoid tactical mishaps, and improve survivability.

Whether these opportunities can be realized depends on whether the information can be presented to the soldier in ways that do not hamper human performance. Of paramount interest is the specific configuration of the physical equipment that will be used to convey this information—particularly the equipment

[1]Packet switching is a method of efficient data transmission whereby an initial message is broken into relatively small units that are routed independently and subsequently reassembled.

that the individual, dismounted soldier will carry. Three related issues that may be equally important are the content of the information to be provided, the format of the messages, and the visual quality of the scene depicted on the screen. Concerns have been raised about the degree of control over the information flow that will be given to the soldier. The problem for system development becomes increasingly serious when it is recognized that issues of format, content, and display design all interact. For example, a particular type of information may call for a format that is not compatible with an otherwise promising display device. Since the land warrior will view the world through the display at night, the quality of the image is critical to his performance.

THE PANEL'S SCOPE AND APPROACH

The panel's work focuses on the compatibility between the characteristics of the proposed equipment and the capabilities and limitations of the target population. Our analysis has been guided by the context in which the helmet-mounted display would be used and the tasks that would be performed by infantry soldiers and squad leaders equipped with the display. Military environments are increasingly varied with respect to both their physical aspects and their task conditions. In addition to warfare, military missions include antiterrorist operations, catastrophe relief, and peacekeeping, among other operations. Equally important are the capabilities and limitations of infantry soldiers as users of the helmet-mounted display. Both system design and training must support their effective use of the system.

One important set of questions raised in this analysis concerns the increased capabilities of the new helmet-mounted display and their effect on individual soldier's ability to think and act independently. Is the Army's intention to empower the soldier to make decisions that are now being made at higher levels? Although such matters of doctrine are beyond the panel's charge, we raise this issue for consideration because it is central to decisions about the ultimate disposition and use of the equipment.

The panel has assessed the scientific evidence regarding the major technical issues, design considerations, and testing approaches associated with the proposed helmet-mounted display. Most critical are the visual and psychomotor factors associated with each design feature. In addition, the stress of the battlefield, including physical sources such as heat, noise, and vibration, as well as cognitive sources such as information overload, complexity, and distraction, affect the soldier's ability to perform with speed and accuracy. And his awareness of his situation, both global and local, is affected in important ways by the wearing of a display capable of providing many types of information in many possible formats. These effects interact in ways that are not yet well understood for the circumstances of an individual dismounted soldier.

The panel proposes an innovative approach to research and testing by the

Army that incorporates critical human factors considerations for the Land Warrior System's helmet-mounted display. This approach is based on a three-stage research and testing strategy that begins in the laboratory, moves to controlled field studies, and culminates in operational testing. It brings the user into the testing and evaluation process earlier, through controlled testing that combines the environmental variations of operational testing with the controlled conditions of laboratory testing.

RECOMMENDATIONS

Evaluation of the effectiveness of helmet-mounted displays involves questions of whether they will help soldiers move, detect, recognize, evaluate, and make correct decisions about objects on the battlefield. Different display technologies possess different attributes that affect a soldier's sensory, perceptual, and cognitive performance. The tasks themselves vary across missions and environments. Trade-off evaluations must take into account the interactions between soldier task demands and the attributes of different devices, because these interactions affect the perception, attention, situation awareness, and workload of the soldiers using the devices. A device that assists a soldier with one task in one environment may impair his performance on a different task or in a different environment. To yield valid predictions about the effectiveness of helmet-mounted displays, the devices must be tested in the laboratory, in controlled field studies, and in realistic field conditions.

After reviewing the available Land Warrior specifications and the existing human factors research findings that apply to those specifications, the panel developed a set of conclusions and recommendations for research and design. They are summarized below and provided in more detail in the final chapter of this book. For some areas, such as the visual properties of the system, there are sufficient data to make concrete suggestions. For other areas, such as the implications of the system for altering workload or situation awareness and the effects of these alterations on performance, additional data are needed before recommendations can be made.

The panel's overarching conclusion is that the proposed monocular system as compared to a binocular system will degrade user performance in the field, and it may also have unacceptable implications for training and selection. As a result, the panel recommends that the Army should proceed in an experimental mode, comparing the positive and negative performance implications of the monocular helmet-mounted display with alternative technologies.[2] Moreover, even if the visual issues are resolved, shifting the infantry soldier's attention away from the

[2]A 1996 report from the General Accounting Office concludes that several human factors issues associated with the Land Warrior System are not yet resolved (U.S. General Accounting Office,

battlefield toward a computer-generated display may compromise situation awareness and increase workload. The panel recommends that, if the display of digital data partially occludes the soldier's view of the environment, then hand-held or wrist-mounted displays should be considered as an alternative to the helmet-mounted display for digital data in order to reduce the likelihood of negatively affecting the soldier's local situation awareness.

Need for an Experimental Approach

The proposed Land Warrior System can be a valuable research tool, if the Army takes an experimental approach to its development. If put into the hands of users in a experimental mode, the Army can establish baseline data and threshold values for future developmental efforts. The panel identified eight general areas for which research is needed:

- Understanding the relationship between design attributes, human attributes, and successful performance for the Land Warrior system is needed for effective personnel selection and training.
- Threshold values are needed for screen clutter, gray scale, limits of spatial and temporal resolution, the impact of visual acuity differences in soldiers, short-term memory limits in processing the information, individual susceptibility to various levels of incapacitation associated with visual rivalry, depth cues, field of view (versus resolution) values, delivery modality preferences and trade-offs, and the impact of attentional narrowing.
- There are several questions associated with the small field of view provided by the proposed helmet-mounted display that require additional study: How are successive glimpses of the display organized by the visual system into a single perceptual image? How much structural overlap is required? Over how much delay? Over how many shifts in view? How is this information combined with outside information? What is the overall impact of the system on soldier situation awareness, when used in combination with other equipment and information in the battlefield environment?
- The question of rifle stabilization when the rifle is extended, aimed, and fired from a protected position has not been adequately addressed. In the current weapon system, the sighting device may be considered accurate, but the aiming variance associated with holding the unsupported weapon stable is large, particu-

---

1996). We concur with the general thrust of the report, although our view is that sufficient specimens of the Land Warrior System—including the helmet-mounted display subsystem should be acquired for research purposes and compared with alternative technologies. Evaluating such specimens in a realistic setting should help answer the questions raised by the GAO report as well as those raised in this committee's report.

larly under conditions of sustained performance. Tests of the accuracy of the aiming of the rifle should be included in the field research program.

• Additional data are needed on the relationship between physical sources of stress on the electronic battlefield (e.g., temperature, vibration, noise) and performance. One important area concerns the potential effects of vibration and small shifts in helmet alignment (caused by walking or more violent motion) on the effective use of the helmet-mounted display. Another area is the combined effects of different levels of physical and mental workload on soldier performance over extended periods of time.

• Comparisons of the Land Warrior monocular system should be made with the existing biocular night vision goggles worn by infantry soldiers and the binocular night vision goggles worn by aviators to determine if significant differences in task performance exist.

• In a given battle situation, the pace of engagement would allow only seconds for reading a display. Determination should be made of what data are critical to show visually to the infantry soldier during combat versus presenting them to the ear.

• Testing and evaluation should be undertaken to ensure that the weight and distribution of the helmet and the display do not interfere with the ability of the soldier to move freely and aim his weapon accurately.

Recommended Design Guidelines

The panel recommends a series of design guidelines for maximizing the soldier's situation awareness and facilitating his ability to process information efficiently. The most important of these include:

- Minimize the degree to which the display is a physical barrier to acquiring information about the environment.
- Provide integrated information in task-oriented sequence, minimizing extraneous information and memory requirements.
- Use graphics that have been well learned by the soldier.
- Simplify the presentation of data entry and system control options.

Much of the technology being considered for the Land Warrior program—and the helmet-mounted display subsystem in particular—is in the early stages of development and the effects of this technology on human perceptual and cognitive performance are, as yet, not well understood. The primary criteria must be based on the soldier's ability to survive and perform. Equipment choices that impede mobility or local situation awareness or both are poor choices regardless of cost.

1

The Military Environment

The conditions under which military missions are or are likely to be conducted are increasingly varied with respect to the physical conditions, the number of tasks, and task complexity. Soldiers in today's Army are involved not only in war, but also in antiterrorist operations, catastrophe relief, and peacekeeping. Each of these missions could require different equipment configurations to achieve optimal soldier performance. And each entails sources of environmental stress that have implications for the soldier's equipment. This chapter provides a basis for our discussion of the human factors considerations related to the Land Warrior System and the subsystems that constitute the helmet-mounted display, describing the military environment in which the proposed new system and subsystems are expected to be used.

COMBAT SETTING

The abilities of the soldiers and the unit to shoot, move, communicate, and survive are the important measures of combat performance for military units, including infantry squads and platoons. Advanced weapons systems and new technologies offer the possibility of increasing individual and unit performance. Advanced weapons technology linked with rapidly unfolding information technology holds the promise of increased speed, accuracy, and lethality for the Land Warrior of what the Army calls "Force XXI" (U.S. Department of the Army, 1994). These technological advances could make soldiers and their units significantly more effective than they are today. However, realizing this potential can be achieved only if the new technology reduces a soldier's fear of going into combat, diminishes the uncertainty of success in combat, and makes it physically less demanding for soldiers and their units to achieve mission success. The land warrior of tomorrow, equipped with sophisticated system capabilities, will be a more formidable fighter only if the new systems build individual soldier confidence and enhance unit cohesion.

Role of the Infantry Soldier

The mission of the infantry soldier in a war-fighting environment in the 21st century will not be dramatically different from the mission of today's soldier. In a war-fighting sense, tomorrow's land warrior will still be required "to close with and destroy the enemy by means of fire and maneuver to defeat or capture him, or to repel his assault by fire, close combat, and counterattack" (U.S. Department of the Army, 1992). However, the role of tomorrow's land warrior may well be changed in significant ways. In recent years, infantry soldiers have been committed to a series of less than combat employments in Somalia and Haiti. The infantry soldiers of tomorrow could be required to react to an even broader array of noncombat or not-active combat requirements, such as peace enforcement, stabilization and civil support missions, counterinsurgency and counterterrorism, humanitarian relief in a potentially hostile environment, and political interventions. Changing international power balances may increase the frequency of low-intensity conflicts, bring about increased counterterrorism actions, and expand the range of peacekeeping tasks.

In these settings, rapidly changing danger situations, avoidance of civilian casualties, restrictive but variable rules of engagement, and the wide dispersal of small units place increased demands on effective communications, accurate position reporting, and detailed intelligence gathering. In general, such roles place soldiers and small units in an environment in which the enemy threat is often unclear, unit dispersions are potentially much greater, and the appropriate response and battle actions are more uncertain. Such changes add to the task of battle-hardening soldiers and their units to deal with the uncertainty and fear of combat.

Future Infantry Combat

Since the end of World War II, the infantry, like the entire Army, has been prepared to respond to a broad range of combat and combat environments. Infantry soldiers have been trained to meet the challenge of high-intensity nuclear war in Europe; to mid-intensity conflict in Europe, the Middle East, and Korea; and low-intensity conflict in Third World countries around the world. For the past 40 years, the Army has been guided by the presence of a clear threat, which led to prescribed doctrine and to weapon systems and organizations driven by capabilities of the threat.

With the end of the cold war have come fundamental changes in the nature of the threats to U.S. security and to the approach of doctrine and organizations. As a result of threat changes, advances in technology, new weapons, and digital information systems, the infantry soldier of the future may see significant changes in basic unit organization at the squad, platoon, and higher levels. The requirement for new, flexible Army doctrine has been clearly presented (see U.S. Department of the Army, Training and Command Center, 1994). This first look into the future characterizes the new battle dynamics of the Force XXI with new dimensions in battle command, battle space, depth and simultaneous attack, and early entry. These dynamics project decreases in the number of soldiers committed to a mission, decreases in mission duration, increases in individual and unit dispersion, extended engagement ranges, increased speed of maneuver, and more complex combat maneuvers. Increased distances and faster tempos add emphasis to the need to prepare individual soldiers and small units to overcome the fear and uncertainty caused by isolation and rapidly occurring battlefield events.

In the past, the cohesion and fighting confidence within a squad or platoon could be maintained by the close proximity of the unit members, by direct and personal leader contact, by local visual contact, and by the direct reinforcement of hand and voice signals. In critical close combat situations, individual performance and unit confidence were enhanced by drawing even closer together or by more deliberate and controlled movements. The use of tomorrow's Land Warrior will tax a unit's ability to retain its confidence and cohesion in the face of greater individual dispersion and the potentially increased speed of maneuver required just to survive on a more lethal and potentially less defined battlefield.

All of the expected differences in the nature of the battlefield reinforce the requirements for increased soldier performance and for appropriate use of new technologies. These requirements include accurate and almost instantaneous information about the position of friendly and enemy elements; more effective communications, down to and within the infantry squad; more accurate and longer-range day and night engagement systems; and increased soldier protection.

These changes and improvements must be achieved by balancing the sensibilities and capabilities of individual soldiers and small units with the advantages and capabilities of new technologies. The changes must address the issues that new technologies raise with their potential for separating a soldier from his local physical environment and the more direct personal contact of today's battlefield. The use of enhanced technology must also be achieved without placing unmanageable cognitive demands on soldiers that could shorten the duration of time allowed for individual performance, that could retard individual response time, or that could become physically debilitating under combat conditions.

Other challenges also affect the nature of combat for tomorrow's infantry.

The end of forward stationing of U.S forces on overseas bases adds greater importance to the issues of deployability and sustainability. Lower budgets reduce the level of infantry forces; they also reduce training dollars and increase the importance of enhancing soldier effectiveness. These same constraints add emphasis to requirements for joint and combined operations and the need to have U.S. infantry soldiers trained, equipped, and able to interface flexibly with a wide range of organizations and systems. These requirements add to the number of tasks for which soldiers must train to maintain their level of combat readiness; they add, as well, to the range of environmental conditions under which the varied missions may have to be conducted.

In addition to these challenges, future employments and the technical complexity of future weapon systems will increase the number of training tasks required and decrease the time available for training. It is likely that a greater number of future infantry soldiers will be part of the National Guard or Army Reserve force structure. This sharply reduces the time they are available for training, since Guard and Reserve soldiers have significantly less time available to train and maintain readiness. These conditions add new dimensions to the problems of maintaining individual and small unit combat readiness. They add as well to the difficulties of developing physical and psychological readiness.

THE LAND WARRIOR SYSTEM

The Land Warrior System is a major research and development effort by the U.S. Army to equip infantry soldiers for the high-technology battlefield of the future. In addition to improving lethality, command and control, survivability, mobility, and sustainability for individual infantry soldiers, it is being designed to provide vision enhancement (under both daytime and nighttime conditions), secure voice communication, greater protection, reduced load, and adequate support for individual maintenance in the tactical environment. Military planners have conceptualized an ensemble of equipment that includes new protective garments, armaments, and information-processing elements. The visual display component of the information-processing subsystem is envisioned as a flip-down monocular presentation device mounted to one side of the soldier's helmet.

The complex technology and pervasive impact of the proposed Land Warrior System on infantry soldiers raise a number of basic doctrine questions about the autonomy of an individual soldier. Is it a design objective to give tomorrow's land warriors information that could allow them to make more independent decisions about their individual tactical actions? At the level of system design, this issue translates into questions about control prerogatives. For example, should an individual soldier decide when a display will be used or moved out of view? If so, should an individual soldier be able to choose the kind of information being presented? Also, should soldiers be able to call up the latest map overlay whenever they want to see it? The concept of a helmet-mounted display with a flexible set of modes or displays that allow soldiers to view different levels of information presentations has significant potential for cognitive distraction. (The negative and positive effects that multimodal displays can have on immediate individual situation awareness are addressed below.)

Other questions concern the nature of two-way communications in the chain of command. Information acquired by an individual soldier from advanced sensor systems (image intensifier, infrared, video camera, lasers, etc.) is intended for use at battalion and higher levels. Should such information be transmitted directly to higher levels or through each level of the chain of command? Is there an operational trade-off between delays in transmission and the possibility of misunderstanding the message at higher levels or misunderstanding the operational situation if intermediate levels do not process the information? Historically, in demanding combat situations, low-level leaders could and would reduce the flow of higher-level instructions and orders by ignoring or not answering transmissions. Will the Land Warrior communications system cause even greater potential for selective avoidance of communications? How will communications security be addressed? What happens if components of the system fall into enemy hands? How will intelligence updates be processed down to the soldier level? Will the squad leader organize and filter information for uploading? Other questions concern the amount or level of interaction among squad members to correct or update visual data and set priorities for information. Although procedures can be developed to address these questions, the most significant question is whether the Land Warrior System will create a situation of task and information overload for individual soldiers and lower-level leaders. The issue is information management. Is there to be a team-based approach to information management and, if so, how is the information distributed among squad members?

The essential measures of human performance for tomorrow's land warrior are not different from those of today's infantry soldier—speed and accuracy. The time it takes a soldier to execute, with precision, a critical combat task is the measure of battlefield success and survival. In combat situations, survival depends on soldiers' ability to rapidly detect, identify, and successfully engage a hostile enemy before the enemy can successfully engage them or their units. Detection and identification rely on the ability to perceive obstacles in the environment, the layout of the terrain, and the location of the present site within the more general situation. For small unit combat actions, this infantry capability is supported by at least three other essential capabilities: accuracy by unit leaders in knowing and reporting their unit's location; the accuracy and timeliness with which unit leaders know the location of friendly units and elements within the range of the weapon systems they control; and the ability to report enemy locations accurately, rapidly, and stealthily.

The infantry rifle platoons and squads of today, and probably of those tomorrow as well, must be prepared to conduct three basic tactical operations: movement, offense, and defense (U.S. Department of the Army, 1992). All of these operations are conducted by day and night, in conditions from desert heat to winter snows, and in terrain that varies from flatlands to swamps to mountains. Under wartime conditions, all infantry missions are planned for execution on a 24-hour-a-day basis. This places great stress on the need for endurance and physical conditioning.

Critical Battle Tasks

Battle success requires the successful execution of a series of critical individual, collective, and leader tasks. Many of these critical infantry tasks overlap and recur many times across all squad and platoon missions. Table 1-1 highlights the critical Land Warrior battle tasks for an infantry soldier, squad leader, and platoon leader for three of the most significant infantry missions: reconnaissance, attack, and defend. The relative importance of specific tasks will vary, depending on the conduct of a particular mission and the level of intensity of combat. For example, the time available for planning varies with the timing of a mission and whether the conflict is a mid- to high-intensity operation, a low-intensity operation, or an operation other than war. In planning for a move in a mid-intensity environment, there may be days available, but in a hasty attack in the same environment, there may be only minutes. In an operation other than war, a squad leader with a mission to defend a key area and control elements of the civilian population would be very focused on the rules of engagement, identification of potential threats, and ensuring appropriate use of force; in contrast, in a mid-intensity defend mission, the leader would be concerned with planning fires, controlling direct fires, and rapid response with all and any firepower available. These tasks and those for the squad leader are drawn from the Hardman III Analysis of the Land Warrior System (Adkins et al., 1995).

Mission and Task Performance

To understand the human factors effects of the proposed Land Warrior System with its helmet-mounted displays, it is necessary to correlate the critical tasks that infantry soldiers and leaders must accomplish and the functional capabilities of the proposed system. Tables 1-2, 1-3, and 1-4 are a series of matrices of Land Warrior System components associated with their expected battlefield enhancements (lethality, tempo, survivability, and mobility) and with key battle tasks performed within a rifle platoon by infantry soldiers, squad leaders, and platoon leader, respectively. The tables do not indicate whether the component aids a soldier or leader in accomplishing a task but only whether the component potentially affects the task.

Although the tasks of the land warrior and his unit may appear relatively simple in terms of these tables, they are in fact difficult because of the need for precise execution by a soldier (with varying cognitive, physical, and general skill levels) and for rapid coordinated execution by the entire unit under battlefield conditions. For example, during a unit attack, when contact with the enemy is made, accurate and responsive reporting of all information is essential in order to rapidly generate combat power at the point of contact and the appropriate element of enemy's combat power.

The first few seconds after enemy contact determines the fate of the maneuvering fire team. On enemy contact, soldiers might immediately drop to the ground, roll for cover, and return rapid fire in an attempt to suppress the enemy fire, or they may be required to rapidly move out of the kill zone and seek cover. In an instant, soldiers wearing a helmet-mounted display would have to execute a broad range of cognitive processing and physical activity: detect and analyze the threat, its location, and effect; drop and roll and return fire or run at a sprint; or drop to cover and call for fires on the enemy position and, perhaps, return direct fire on the location. If at that moment the Land Warrior System display interferes with the soldiers' immediate situational awareness or hinders their physical movements, all the system's technological capabilities may not matter.

The first minutes after an initial enemy contact determine whether the squad and platoon will be successful in maintaining contact by rapidly increasing rates of fire and executing squad maneuvers to fix the enemy. Knowing friendly positions and the location of the enemy accurately is essential if requests for fire support are to be timely and effective. In the future, because of the potential for increased dispersion and more rapid movement, it will be more imperative and time-sensitive for those in higher-level command and control positions to know the exact location of all friendly forces in order to provide responsive coordinated support and at the same time avoid casualties from friendly fire.

During the conduct of an attack, the physical and neurological responses of a soldier could be critical to his survival. Today's soldiers already have an almost overburdening load of equipment. If the added weight of Land Warrior System components slow them or distract their attention from the local surround, even unintentionally, the outcome could be negative.

The design of the helmet-mounted display must be compatible with the physical responses a soldier has to make and, to be effective, must accommodate the different psychological and neurological capabilities of different soldiers. A flexible display with a selection of modes of information presentation and levels of detail may not be fully compatible with the capabilities of a given soldier or with the actions required in battle. An illustration of a negative impact is the disorientation that an individual soldier would experience if he were moving at night, wearing a helmet-mounted display in a night vision mode with a field of vision of 60 degrees, when suddenly his unit is engaged. He drops to the ground and rolls to cover, switches to the night thermal sight with a field of vision of 9 to 15 degrees, and attempts to determine the direction of enemy fire of which he is totally unsure. At the same time, the impact of a bright flare shuts down his night

TABLE 1-1 Critical Land Warrior Battle Tasks

	Reconnaissance	Attack Mission	Defend Mission
Soldier	Receive squad order	Receive attack order	Prepare position
	Move tactically	Move tactically	Search assigned sector
11B10	Identify ORPs	Recognize obstacles	Detect targets
	Observe/listen for enemy	Detect enemy/target	Report detection
	Conduct reconnaissance	Identify enemy locations	Identify enemy
	Locate objective area	Determine range	Determine range
	Observe enemy activity	Follow rules of engagement	Respond to orders
	Record information	Engage targets	Engage target
	Report enemy information	React to enemy actions	Report status
		Respond to squad order	
		Assault objective	
		Report status	
Squad	Receive platoon order	Receive platoon order	Plan defense
Leader	Evaluate move route	Issue squad order	Plan fires
	Move tactically	Move tactically	Issue order
11B30	Control squad movement	Control squad maneuver	Control defense preparation
&	Land navigate	Land navigate	Prepare squad overlay
11B40	Identify ORPs	Recognize obstacles	Detect targets
	Determine location	Detect enemy/targets	Identify enemy
	Report location	Identify enemy locations	Determine range/location
	Observe/listen for enemy	Determine range	Report enemy status
	Conduct reconnaissance	Enforce ROE	Respond to orders
	Locate objective area	Control direct fires	Engage targets
	Observe enemy activity	Call for indirect fires	Control direct fires
	Record information	React to enemy actions	Call for indirect fires
	Receive enemy information reports	Engage targets	Receive status report

	Report enemy locations	Respond to platoon orders	Redistribute ammunition
	Report situation	Assault objective	Report situation
		Consolidate/reorganize	
		Report situation	
Platoon	Receive company order	Receive company attack order	Receive company defense order
Leader	Plan reconnaissance	Plan maneuver	Plan defense
	Issue reconnaissance order	Preplan fires	Plan fire support
(11)	Move tactically	Issue platoon order	Coordinate left and right
	Control platoon movement	Move tactically	Issue platoon order
	Land navigate	Control platoon maneuver	Control defense preparation
	Identify ORPs	Land navigate	Prepare platoon overlay
	Determine location	Control direct fires	Report enemy contact
	Receive element reports	Call for indirect fires	Control defense actions
	Report location	Coordinate left and right	Call for indirect fires
	Observe/listen for enemy	Respond to company orders	Receive status reports
	Direct reconnaissance	Assault objective	Report defense status
	Locate objective areas	Direct consolidation	Respond to orders
	Receive reconnaissance reports	Reorganize/resupply	Reorganize/resupply
	Record information	Receive squad status report	Report situation
	Report situation	Report situation	

TABLE 1-2 Land Warrior System Components and Their Impact on the Soldier's Tasks

	Lethality					Tempo				Survivability					Mobility		
Level and Task	Thermal Weapon Sight	M4/M16 CCO	IR Aiming Light	Laser Range Finder	Laser Detection Device	Half Duplex Radio	Computer	Video Camera	C4I Control Display	Thermal Weapon Sight	Laser Detection Device	BLEPS Protection System	PASGT	Ground Position System	Integrated Headgear	Battery	Ground Position System
Receive order						x	x		x						x	x	x
Move tactically											x	x	x		x		
Detect targets	x	x		x	x					x	x	x	x	x	x	x	
Identify enemy	x	x	x	x	x					x	x			x	x	x	
Determine range				x				x						x	x	x	x
Engage target	x	x	x	x	x					x		x	x		x		
Respond to orders			x			x	x	x	x						x		x
Report status						x	x	x	x								x

vision devices, and the soldier may be disoriented to the point of not being able to make a response.

At the squad level in an attack, the situation may be different. The critical tasks that the helmet-mounted display supports include controlling direct fires, maneuver, and coordinating indirect fire support. A successful squad attack requires a squad leader to rapidly adjust the direct fires of his unit; accurately report the situation, including the location of his unit; make timely calls of fire support; and control the maneuver of his squad. All of these actions must occur under enemy fire and amidst confusion and uncertainty.

Defensive operations may have a different set of effects on the design of the Land Warrior System. A defense can be characterized by periods of intense preparation, followed by irregular periods of boredom and observation waiting for enemy activity, and then the overwhelming impact of an enemy closing in on the defensive position. Preparation for defense is an intense, day-night, physically demanding, individual, and team effort. After intense preparation, leaders and soldiers are often physically and psychologically stretched. The impact of the helmet-mounted display could be a stress multiplier if its design causes a false sense of capability or a lack of local situation awareness that leads to ineffective soldier responses.

In defensive operations, leaders fight by ranging targets, setting priorities for targets—what and when to fire—controlling fires to destroy the nearest or most dangerous, and shifting direct, indirect, and indirect supporting fires. These actions require constant communications and rapid information flow with accurate reports when soldiers and leaders may not be at peak physical and psychological levels. The design of the Land Warrior System and the helmet-mounted display have to meet these challenges to achieve their expected potential. Using a helmet-mounted display that has a selection of modes effectively means that the leaders and soldiers must have a way of determining what information is needed, so that everyone has the correct situation awareness at the same critical time. In defensive operations, the system must allow for the individual capabilities and differences of the soldiers using it.

Psychological Considerations and Team Building

Although any given task may not be particularly demanding in a benign environment, the collective squad or platoon tasks associated with shooting, moving, and communicating are performed under conditions of great stress. The sources of stress include the nature of the mission, resource availability, perceived risk, information uncertainty, physical demands, fatigue, time, environmental conditions associated with weather and terrain, and surprise. As a result of these stresses, a soldier's state of arousal varies from prolonged boredom to short periods of stark terror. Much of the cognitive workload is the result not only of the tasks to be performed, but also of the standards to be achieved and the

TABLE 1-3 Land Warrior System Components and Their Impact on the Squad Leader's Tasks

	Lethality					Tempo				Survivability					Mobility		
Level and Task	Thermal Weapon Sight	M4/M16 CCO	IR Aiming Light	Laser Range Finder	Laser Detection Device	Half Duplex Radio	Computer	Video Camera	C4I Control Display	Thermal Weapon Sight	Laser Detection Device	BLEPS Protection System	PASGT	Ground Position System	Integrated Headgear	Battery	Ground Position System
Receive order						x	x	x	x					x	x	x	x
Issue orders						x	x	x	x								
Move tactically			x	x					x		x	x		x	x		x

Control move				x		x	x	x	x			x	x		x
Land navigate												x	x	x	x
Identify targets	x			x	x				x	x	x	x			
Engage targets	x	x	x	x						x	x		x		
Respond to orders			x			x	x	x	x			x	x	x	x
Report location						x	x		x			x	x	x	x
Central direct fire			x	x	x	x	x	x	x			x	x	x	x
Call indirect fires				x		x	x		x			x	x	x	x
Receive report					x	x	x	x	x			x	x		
Report situation						x	x	x	x			x	x		

TABLE 1-4 Land Warrior System Components and Their Impact on the Platoon Leader's Tasks

	Lethality					Tempo				Survivability					Mobility		
Level and Task	Thermal Weapon Sight	M4/M16 CCO	IR Aiming Light	Laser Range Finder	Laser Detection Device	Half Duplex Radio	Computer	Video Camera	C4I Control Display	Thermal Weapon Sight	Laser Detection Device	BLEPS Protection System	PASGT	Ground Position System	Integrated Headgear	Battery	Ground Position System
Receive orders				x		x	x	x	x					x	x	x	x
Plan				x		x	x	x	x					x	x	x	x
Issue orders						x	x	x	x					x	x	x	x
Coordinate				x		x	x	x	x		x			x	x	x	x
Land navigate									x		x	x	x	x	x	x	x
Control move	x		x	x	x	x	x	x	x		x		x	x	x	x	x
Control fire	x		x	x		x	x		x	x				x	x	x	x
Call indirect fires						x			x		x			x	x		x
Receive report						x	x	x	x					x	x	x	x
Report status						x	x	x	x					x	x	x	x

conditions under which the tasks have to be performed. The challenge for system design is to avoid unanticipated negative consequences that will impair confidence, cohesion, commitment, and communication. Losing or degrading the quality of information a soldier is currently receiving under daylight conditions and which he perceives to be critical could affect how and under what conditions he will use the system.

Success in combat requires that infantry soldiers control their fear and behave in a predictable manner, no matter how tired they are or how uncertain the situation is. In order to elicit this behavior, commanders attempt to build high-performing, cohesive, and confident units by clearly defining roles and jobs, keeping soldiers informed, conducting realistic and demanding individual and group training, and providing strong leadership and competent supervision. The introduction of new technologies and capabilities adds, deletes, or modifies tasks; affects how tasks are performed; changes job requirements; enhances or degrades confidence and cohesion; changes training requirements; and affects the acquisition and retention of personnel. The Land Warrior System is not an exception: it will affect how leaders define soldiers' roles and jobs and how they communicate with and train and lead their units.

Military commanders have known for centuries that fear is easier to cope with in groups than in isolation. Building confidence at the individual and unit levels is essential to overcoming fear. As a result, "buddy" systems have existed for many years. Extending the buddy system concept through the fire team, squad, platoon, and company levels builds unit cohesion and instills trust and interdependence in a larger group. Technology can facilitate or hinder unit cohesion, depending on how it is designed and used. Communication technology allows soldiers to feel a part of a unit despite geographical separation. In contrast, a display that reduces vision in daylight, induces motion sickness, increases workload and reading time, causes discomfort, or requires constant adjustment will negatively affect individual confidence and unit cohesion.

Individual and collective training is conducted to improve confidence, overcome fear, and enhance performance reliability. With sufficient training, soldiers know that they can execute assigned tasks under all conditions and achieve or exceed the expected standards. Training helps the soldiers gain confidence in their roles within the unit for any specified mission. The military structure assists in establishing, maintaining, and reinforcing clearly defined roles. Role clarity fosters interdependence and trust as well as confidence. The reporting procedures associated with new technologies could undermine this structure and erode soldier confidence in unit leaders if they are used to bypass the structure.

Radios have been used reliably at the platoon level and higher and for many years. Although platoon and squad radios have existed, they have not always been reliable, and squad members do not carry radios. As a result, communication below the platoon level has historically relied on visual or auditory contact with another member of the unit. But hedgerows, dense jungles, mountainous terrain, and urban environments make it difficult to maintain visual and auditory contact. The effects of poor communication on maintaining a cohesive fighting force, avoiding fratricide, and shortening battle execution time are obvious.

Communications in the military usually include orders and instructions flowing from the headquarters to soldiers, and information about the battle situation flowing from small unit leaders to the headquarters. As communication technology has improved, so has the demand for information at the headquarters level. When these demands from headquarters are not synchronized with the work ongoing in units that are engaged in battle, the communication process itself can reduce battle effectiveness by diverting leader attention and by adding stress to an already overstressed situation. Personnel who are overworked will shed tasks until they achieve a level they can manage. Reporting takes time and can be a major workload factor; technology should be used to reduce this workload, not to increase it.

The natural human tendency is to want more information, not less. Each information requirement should be questioned as to who needs it, when they need it, under what conditions they need it, and how they will use it. In terms of display, more information frequently leads to more complex screen designs, navigation menus, function keys, and other control mechanisms. Ultimately this could further contribute to reduced situation awareness, with the unintended consequence of decreased survivability.

In 1959 General Bruce C. Clarke stated: "The truth is that the most expensive weapon that technology can produce is worth not an iota more than the skill and will of the man who uses it." The challenge for engineering design is to use technology to improve a soldier's skill and will. History is replete with examples of technological advances that were introduced into the military before they were introduced into society in general and did not have the desired effect (Guilmartin and Jacobowitz, 1984).

Decision Making

The use of new technologies to improve communication, provide rapid access to data on enemy and friendly positions, improve soldier protection, and provide more accurate daytime and nighttime engagement systems implies improved combat effectiveness. Command and control at the battalion task force and higher levels should improve; it is not evident, however, that the use of new technology will help the squad leader during wartime operations as much as it does higher echelons of command. Although command and control is clearly outside the scope of this study, we believe that an important consideration in evaluating the helmet-mounted display design is how well it helps or hinders the squad leader and the platoon leader in doing their jobs. To that end, it is important to understand decision making at the squad level and its impact on workload and squad performance.

Decision making consists of choosing between alternative courses of action (Buck and Kantrowitz, 1983). The quality of a decision is heavily dependent on the information available to the decision maker. Consequently, the processes of communication cannot be separated from the processes of decision making. The first step in any decision-making process is the recognition that some action must be taken. In scientific terms, the decision maker observes that a salient condition is not within some bounds of tolerance. This could be as simple as the recognition of hunger, with the prime decision being that of taking action to seek food. In a combat setting, the typical instance might be the detection of an immediate threat for which the action could be to take cover or attack. However, in both instances, the initial observation plus action selection is only the preliminary stage to what can become a long series of incremental decisions. In the combat situation, the sequence would include a determination of optional means to counter the threat, calculations of the risk factor and the likely outcome from each such means, and selection of a course of action intended to suppress or eliminate the threat.

It is clear that, in combat, leaders are confronted with anything but trivial decisions. They are deeply engaged in information collection and absorption, the formulation of mental representations of external conditions (situation awareness), the assessment of probabilities, and the weighing of the options to give a net outcome value. In this regard, the complexity of the problem can easily exceed the mental capacities of the decision maker and lead, in turn, to the need for provision of external support, such as decision aids of one kind or another.

Over the years the Army has evolved a number of techniques for managing the workload and communication at the squad level. Generally these have involved the use of standard operating procedures, battle planning, rehearsals, battle drills, crew drills, and extensive training under realistic conditions. The result has been to reduce the need for communication and increase the speed of battle execution. Essentially the squad leader and the squad learned to quickly identify a situation and react to it with a battle drill rather than have to think about it. The role of the squad leader was to lead by example. He was not required to creatively solve new problems without assistance from his platoon leader or platoon sergeant.

The helmet-mounted display has the potential for significantly increasing the workload of the squad leader by adding both information-gathering and decision-making tasks to his existing job. Some squad leaders read and write at a minimally acceptable level, mastering other techniques for learning. The helmet-mounted display will require the squad leader to read and write more than he has historically. Under these conditions, it is anticipated that the squad leader will be slow and deliberate and will gain in speed only with training; as a result, his initial workload levels will be very high. When these new tasks are combined with the time stress associated with battle, many squad leaders will be overloaded.

Other concerns are raised by the fact that the helmet-mounted display will separate the squad leader from his natural view of the environment. The extent of this separation is not known, nor is the impact of it fully understood in terms of performance. Viewing the world through the display at night, rather than reading the display to extract or enter data, removes the squad leader from the environment in different ways.

It may be that the helmet-mounted display will be of greatest assistance to the squad leader in operations other than war. Such operations pose a unique set of challenges, which require exercising a different degree of control over the use of military force than conventional wartime operations, with relatively clear rules of engagement. Whether or not the technology will improve battlefield effectiveness will be determined in part by the quality and timeliness of the decisions made by the squad leader and the chain of command.

There are a number of important questions that should be addressed by the Army regarding the implications of the helmet-mounted display for information processing and decision making:

- How do decision-making factors vary in order and significance from situation to situation, and what does that mean for the design of the helmet-mounted display?
- To what extent does local situation awareness[1] affect decision-making at the squad level?
- To what extent does global situation awareness[2] affect decision-making at the squad level?
- How should presentation formats be designed for decision making under stress (defined as decisions that have to be made in less than a minute with incomplete information)?
- To what extent do poor resolution, field of view, and depth perception affect decision making?
- How will decision-making aids be designed to support the squad leader?
- What decision-making training is currently provided?
- To what extent does leadership contribute to good decision-making?

CONCLUSIONS

The operating environment of today's infantry soldier is varied, complex, and demanding. The environment of tomorrow's land warrior may well be even more varied, more complex, and more demanding. The design issues for the

---

[1]Local and global situation awareness are defined in Table 3-1. Local is characterized by target identification, target location, terrain and object distance, and cueing of a hostile presence.

[2]Global situation awareness is characterized by location of self, location movement of other units, commands and directions from headquarters, and navigation information (see list in Table 3-1).

helmet-mounted display of Land Warrior System that are raised by this discussion of the future infantry battlefield can be summarized in four key themes.

First, what critical doctrinal and employment priorities must or should the Land Warrior System and the head-mounted display meet? Force XXI projections identify changes in the scope of future infantry tasks and present a view of an expanded range of future infantry missions. These projections for tomorrow's land warrior highlight the need for a more clearly defined Land Warrior System doctrine. For example, is the design objective to give tomorrow's land warriors information that could allow them to make more independent decisions about their individual tactical actions? An integrated battlefield information and weapon system incorporating the latest electronic technology cannot be effectively achieved without clear employment concepts. Developing the design of the helmet-mounted display requires technical trade-offs, not only in achievable technology but also in human performance. No one set of trade-offs will optimize human performance under all conditions. The most advanced technical capabilities will not improve human performance unless the system is reliable and builds confidence.

Second, what are the limitations of the cognitive, psychological, and physical capabilities of the infantry soldiers in the active force, the National Guard, and the Army Reserve who will use the Land Warrior System? In order to achieve the expected enhancements in human performance, the Land Warrior System and its head-mounted display must be designed to support the mental, psychological, and physical characteristics of tomorrow's land warrior. Complex electronic displays and battle sights will not achieve the Army's goal if they provide overwhelming levels of information that a soldier may not need and may not be able to process effectively under critical battlefield conditions.

Physical ergonomics also plays a critical role in achieving the success of Force XXI Land Warrior. There could be no more disastrous scenario for advanced battlefield technology than the prospect of squads of soldiers so engrossed in making the equipment work that they become easy targets for the enemy. In contrast, a well-designed, effective information system could not only improve efficiency but also could mitigate effects of stress, which have been a traditional barrier to mission effectiveness in conflict situations. It is important to understand the mechanisms of stress and how they interact with the proposed technology. How to suppress the vibration in a head-mounted display that is caused by walking is also a question for research and design.

Third, what are the significant implications of the Land Warrior System for testing and evaluation? The desire to rapidly develop and achieve a Land Warrior System that applies significant technological advantages to the infantry battlefield requires the development of a comprehensive and thoughtful test and evaluation strategy. The rapid advances made in computer and electronic systems can sometimes lead to expectations about capabilities that are not achievable, do not work, and are not supportable under any realistic battlefield or employment conditions. The Land Warrior System and head-mounted display must be systematically tested in a range of realistic battle conditions, not only to aid the development and refinement of employment doctrine, but also to validate technical capabilities, system performance, and development priorities. Given the lack of information about human performance with helmet-mounted displays in which the body is in motion, a comprehensive test program is needed that considers helmet stability, helmet display capabilities, perceptual understanding and comprehension, and variance associated with individual differences.

Fourth, what are the critical issues in selection, retention, and training of future infantry soldiers? The area of training requires focused attention. The design of the Land Warrior System and its head-mounted display not only must be logistically supportable in terms of reliability, availability, and maintainability, but must also result in a system that soldiers can be taught to use effectively, efficiently, and confidently. If the complexities of the display and the system require extraordinary personnel selection or training demands that do not allow the development and maintenance of combat proficiency in infantry soldiers, then their combat potential could be lost.

2

The Infantry Population

The Land Warrior System has the potential to significantly alter traditional infantry roles, functions, relationships, and employment concepts. Such changes will affect both personnel selection and soldier performance. In this chapter we examine the criteria used by the Army to enlist infantry soldiers, the functions of infantry training, and the implications of infantry cognitive capabilities and anthropometrics for design of the Land Warrior System. We describe minimum requirements rather than the average because, ultimately, the designer's equipment must be usable for all who qualify for enlistment.

PERSONNEL SELECTION

Soldiers are recruited from the general population in accordance with prescribed cognitive, medical, and physical standards (height, weight, and strength).[1] In fiscal 1994 the Army enlisted 68,000 recruits, of whom approximately 63,000 did not have prior service experience. About 10,000 were assigned to the infantry. The Army uses high school degree and the Armed Forces Qualification Test (AFQT), which is a subset of the Armed Services Vocational Aptitude Battery (ASVAB), as criteria for selection. As a matter of policy, the Army can change the distribution of quality within the force by changing these selection criteria. We describe below the infantry population in terms of the selection criteria currently in use.

---

[1]At one time, the Army administered a strength test (MEPSCAT) to screen recruits; this is no longer done. The minimum height requirement is 60 inches, and the maximum is 80 inches.

Cognitive Entry Requirements

The ASVAB is a basic tool for recruitment and selection in the military. It is composed of 10 subtests:

1. General science (GS),
2. Arithmetic reasoning (AR),
3. Word knowledge (WK),
4. Paragraph comprehension (PC),
5. Numerical operations (NO),
6. Coding speed (CS),
7. Auto shop (AS),
8. Mathematics knowledge (MK),
9. Mechanical comprehension (MC), and
10. Electronics information (EI).

Verbal ability (VE), which is considered an eleventh subtest, is a composite of word knowledge (WK) and paragraph comprehension (PC). These subtests are combined in various ways to form the Armed Forces Qualification Test, which is used for enlistment screening and job assignment. Each service establishes its own composites of the ASVAB subtests to satisfy job structure and mission requirements. Applicants are classified in a military occupation by using the composites to predict success in initial occupational training schools.

The Army composites from the ASVAB subtests correspond to 10 aptitude areas:

1. Combat (CO),
2. Clerical (CL),
3. Field Artillery (FA),
4. General Maintenance (GM),
5. Motor Maintenance (MM),
6. Operator and Food (OF),
7. Electronics (EL),
8. Surveillance and Communications (SC),
9. Skilled Technical (ST), and
10. General Technical (GT).

Scores on the AFQT range from 0 to 100. The average is 50 and the minimum acceptable score is 10. Test results are divided into five test score categories: I, II, III, IV, and V, and III is further divided into IIIA and IIIB. Categories I through IIIA represent the upper half of the recruit population. Anyone scoring in category V is prohibited by law from entering the armed services; federal law also restricts the number of category IV recruits to no more than 20 percent of the

TABLE 2-1 Distribution of Scores on the Armed Forces Qualification Test (AFQT)

AFQT Category	Scores	Reading Grade Level	General Ability	Percent of U.S. population
I	93-99	12.7-12.9	Very high	8
II	65-92	10.6-12.6	High	28
IIIA	50-64	9.3-10.5	Average (top)	17
IIIB	31-49	8.1-9.2	Average (bottom)	17
IV	10-30	6.6-8.0	Low	21

annual enlistees. The AFQT categories and what they mean in terms of reading grade level and general ability are shown in Table 2-1.

The Army sets quality targets both for overall enlistments and for each of its career management fields. In 1994 the Army-wide AFQT targets were 67 percent I-IIIA, 31 percent IIIB, and 2 percent IV. For the infantry, the targets were 64 percent I-IIIA, 33 percent IIIB, and 3 percent IV; the infantry actually achieved approximately 68 percent I-IIIA, 28 percent IIIB, and 4 percent IV. The National Guard is not required by law to meet the category IV standard that the active component is required to meet, yet over the last 3 years it has voluntarily met the 2 percent category IV goal.

Although a high school degree is used as a selection tool (it is a good predictor of the likelihood of enlistment completion), it is not technically a requirement for enlistment. The competition is such, however, that for all practical purposes it serves as a requirement. In 1982, Congress placed a ceiling of 35 percent on male Army enlistees with no prior service and no high school diploma. In 1994, the Army achieved its goal of 95 percent high school graduates. About 85 percent of the Army National Guard recruits have high school diplomas.

Results from the 1992 Youth Attitudinal Tracking Survey (Lehnus, 1994) show declining enlistment rates. As recruiting becomes more difficult, quality may be traded off by adjusting the selection criteria in order to meet goals for accession numbers. Of course, in the private sector the demand for the best and the brightest will continue to rise as well. As the economy and job opportunities improve, the competition for high-quality personnel can be expected to increase.

Although it has been rumored that the force is getting older, the facts provided by the Army do not support this notion. The average age of the active force is 20.4 years and has been relatively constant over the last 10 years; no data were gathered on the Army Reserve or the National Guard (information provided by the Office of the Deputy Chief of Staff for Personnel, HQDA). The number of soldiers for whom English is a second language is likely to increase, according to current and projected immigration trends.

It is important to note that, although a high school diploma and ASVAB scores are useful measures of quality and have been shown to correlate with performance in military jobs (Wigdor and Green, 1991; Green and Mavor, 1994), additional screening criteria are needed for selecting soldiers to perform infantry tasks. These criteria and the tools to measure them become even more important as the nature of the tasks become more complex and information-intensive, as they will with the Land Warrior System. Specifically, it is essential to consider the human abilities that will affect soldier performance with the Land Warrior System. Once these are defined, existing selection tests can be identified or new ones developed for measuring these abilities.

One effort to examine the abilities required by different tasks was a 1982 Army study using Fleishman's (1975) human performance taxonomy to determine the degree of difficulty associated with performing tasks in 20 new systems (U.S. Department of the Army, 1982). Known as the "Complexity Study," this study was successful in predicting performance and has been validated through staff studies conducted at the Pentagon. Research that builds on these results should be vigorously pursued by the Army. The taxonomy proposed by Fleishman is presented in Table 2-2 as a guide to identifying useful criteria. The challenge is to identify the relationships between the land warrior helmet-mounted display characteristics and the human attributes that correspond to successfully use the system. Relationships between human attributes and situation awareness are presented in Chapter 3.

Infantry Requirements

The physical requirements for the infantry encompass those for all Army personnel. In addition to the Army entrance requirements, the following specific requirements have been established for the infantry, which is open only to men (U.S. Department of the Army, 1994a):

1. A physical demands rating of very heavy,
2. A specific physical and medical profile,
3. Color discrimination of red and green,
4. Correctable vision of 20/20 in one eye and 20/100 in the other eye,
5. A minimum score of 90 on ASVAB aptitude test for combat, and
6. Formal training under the auspices of the Infantry School (completion of 11B course).

The physical demands for infantry soldiers include the following:

- Occasionally raise and carry 160-pound persons on one's back,
- Frequently perform all other tasks while carrying a minimum of 65 pounds, evenly distributed over the entire body,
- Frequently walk, run, crawl, and climb over varying terrain for a distance of up to 25 miles,
- Frequently give oral commands outside at distances up to 50 meters,
- Be able to hear oral commands outside at distances up to 50 meters,
- Occasionally climb a rope a distance of up to 30 feet,
- Frequently throw 1-pound object 40 meters, and
- Frequently visually identify vehicles, equipment, and individuals at long distances.

The specific physical and medical profile refers to functional capacity to perform (as determined by medical personnel) in six areas:

- Physical capacity: good muscular development with the ability to perform maximum effort for indefinite periods;
- Upper extremities: no loss of digits or limitation of motion; no demonstrable abnormality; able to do hand-to-hand fighting;

TABLE 2-2 Fleishman's Human Performance Taxonomy

Cognitive	Physical	Psychomotor
Verbal comprehension	Static strength	Choice reaction time
Verbal expression	Explosive strength	Reaction time
Idea fluency	Dynamic strength	Speed of limb movement
Originality	Stamina	Wrist-finger speed
Memorization	Extent flexibility	Multi-limb coordination
Problem sensitivity	Dynamic flexibility	Finger dexterity
Mathematical reasoning	Gross body equilibrium	Manual dexterity
Number facility	Gross body coordination	Arm-hand steadiness
Deductive reasoning		Control precision
Information ordering		Rate control
Category flexibility		
Spatial orientation		
Visualization		
Speed of closure		
Flexibility of closure		
Selective attention		
Time sharing		
Perceptual speed		

• Lower extremities: no loss of digits or limitation of motion; no demonstrable abnormality; be capable of performing long marches, standing for very long periods;

• Hearing—ears: audiometer average level of six readings (three per ear) at 500, 1,000, 2,000 Hz or not more than 30 dB, with no individual level greater than 35 dB at these frequencies and level not more than 55 dB at 4,000 Hz; or audiometer level of 30 dB at 500 Hz, 25 dB at 1,000 and 2,000 Hz, and 35 Db at 4,000 Hz in better ear (poorer ear may be deaf);

• Vision—eyes: distant visual acuity correctable to 20/40-20/70, 20/30-20/100, or 20/20-20/400; and

• Psychiatric: no psychiatric pathology; may have history of a transient personality disorder.

Accepting less than perfect vision requires that the Land Warrior System must accommodate soldiers who wear glasses. For older soldiers who wear bifocal lenses, this challenge is compounded; optical inserts in the chemical protective mask are not bifocal.

Some infantry positions require more use of tactical information systems than others. An example is the special forces sergeant. Individuals who enter this career field do so after they are in the Army. In addition to the above requirements, candidates for this rank and status (which is also closed to women) must meet the following:

- A minimum score of 110 in general technical aptitude and 100 in combat aptitude;
- A secret-level security clearance,
- Completion of a formal Special Forces Qualification Course (SFQC),
- Other requirements listed in Army Regulation 614-200, and
- U.S. citizenship.

The position of ranger is also likely to make heavy use of complex tactical information, although there is no separate career field for it. Army personnel assigned to ranger battalions are also selected after they are in the Army and must meet very demanding requirements.

TRAINING

Training is designed to accomplish many purposes, including acquiring technical, leadership, interpersonal, and conceptual skills; improving and sustaining proficiency from individual to enterprise levels; and inculcating values and beliefs. Army training addresses both individuals and groups. Individual training teaches soldiers the basic skills required for individual survival and job performance on the battlefield. Group training prepares soldiers to perform tasks as part of a team. Most individual and some group tasks are taught as part of basic and advanced individual training, whereas some individual and most group training is conducted in units. In general, training time demands exceed available time to train, so unit commanders set priorities for training time on the basis of their missions' essential task list and assessments of training needs.

Demanding and realistic training builds confidence and cohesion. Resource limitations often reduce the amount of training that is provided as part of fielding new equipment. Fielding the Land Warrior head-mounted display may be particularly challenging. Although the fielding teams do a superb job, they only do it once. The unit assumes responsibility for training new personnel that come as a result of normal attrition. As the original personnel and trainers depart the unit, new trainers take their place without the benefit of having the new equipment fielding training. This attrition process combined with the lack of sustainment training results in a loss of knowledge and deterioration in the quality of training over time. This is offset when the new equipment is used in the Army's basic or advanced individual training, such as tanks, artillery, etc. Training on smaller, more numerous items of equipment often takes place within the unit. Embedded training is one means of preventing the deterioration in the quality of the training that may take place.

Implications of the Land Warrior System

Work by the Army Research Institute has shown that cognitive tasks involving the accurate recall of subject matter with little or no behavioral component (e.g., a pilot preflight checklist) have high rates of skill decay and require more sustained training than other tasks to maintain proficiency. The Land Warrior System will have a large number of this class of cognitive tasks at the leader levels. To the extent possible, the number of these tasks should be minimized.

One of the implied outcomes of the Land Warrior System is to increase the speed at which the soldier thinks and acts. In order to achieve this, soldiers at various levels must: (1) Respond to data: give attention to detail, perceive form, recognize and identify patterns, recall rules, and comprehend their environment. (2) Take action based on data: perform quantitative analysis, reason verbally, assess given situations, formulate concepts, plan, and make decisions. (3) Create data: make inferences, formulate and validate hypotheses, and solve problems. All these things must be done to some extent by soldiers at the front.

Historically, soldiers were taught battle drills at the individual, crew, fire team, squad, and platoon levels in order to teach them to recognize a situation and react to it. Furthermore, the squad leader leads by example, which makes it easier for soldiers to see what he was doing and follow his example. These techniques have proven successful in improving reaction time and reducing the time needed for problem solving. Learning standard procedures and drills will continue to assist the speed at which the soldier thinks and acts.

In operations other than war, soldiers find themselves in unique situations for which battle drills and standard procedures have not been developed. In providing soldiers with a common perception of the battlefield or other common contextual framework, the helmet-mounted display will facilitate actions that are based on responding to data or taking action based on data. The extent to which it will speed thinking and acting when the soldier has to create data is not so evident. The challenge is to design a training system that will facilitate this outcome.

The Land Warrior System will provide new capabilities to the infantry squad; however, the major return on investment is dependent on soldiers doing things differently than they do today. How well the soldier can use these new capabilities will be a function of training. The system will affect how the soldier sees and interprets the visual scene. The changes in the viewing world will be significantly different from what the soldier learned through life experiences. Loss of depth information and resolution, alternative fields of view that are narrower than normal vision, and a dynamic environment will present new challenges associated with object recognition, balance, orientation, and movement. In other words, the soldier's ability to detect threats, move across terrain, and maintain local situation awareness will be affected. Training will be needed to alleviate some of these concerns.

The squad leader will face a number of challenges that, although not new in concept, will be new in application and may have training implications: workload management, maintaining 360 degree observation during movement, actions on contact given possible disorientation and changing field of views, maintaining vigilance during slow-tempo operations (overcoming a false sense of capability), squad previsualization training to reduce cognitive workload and attention demands.

With regard to the helmet-mounted display, individual training may be required in a number of areas (although the following list is not exhaustive):

- Equipment operation and adjustment,
- Symbology training,
- Map reading,
- Visual scene cues and interpretation,
- Alternative techniques for viewing the scene and overcoming attentional narrowing,
- Data interpretation,
- Attention switching between the eye viewing the display and the "ambient eye,"
- Object recognition training in the absence of depth cues, color, shading, texture gradient, poor resolution, etc.,
- Engagement techniques with various sensors,
- Movement techniques,

- Computer skills associated with application software,
- Maintenance skills to include field expedient techniques, and
- First aid for motion sickness, blindness, and other temporary maladies associated with system use.

Requirements for individual training will be much greater when other systems associated with the Land Warrior System are included, such as the intrasquad voice and data communications, integrated computer/radio/global positioning system with digital maps and overlays, combat identification, multithreat warning devices, medical monitoring, objective individual combat weapon, a thermal/laser aiming device, a microclimate cooling system, integrated/modular clothing and equipment, and an improved load carriage system.

Training aids may be embedded into the Land Warrior System, provided as stand-alone devices, some combination of the these two, or, in the worse case, not provided at all. Although embedded training systems can help alleviate many problems, the computer hardware, space, power, system architecture, and additional weight requirements must be considered as part of the total process of system design. Some specific hardware and software features associated with installing and operating an embedded training system include processing speed, access rate, memory, display capabilities, communications, interfaces, menu structures, and the capabilities for collection of performance data. Planning for such a system must include an analysis of the conditions of use (e.g., conducting embedded training for one platoon while another is conducting operations in a combat zone). More significant will be the cost associated with designing and acquiring a fully embedded training system. Embedded training is relatively inexpensive from a hardware point of view, but it is relatively expensive to develop courseware programs. As a result, only small amounts of training courseware are available, and its compatibility with the Land Warrior System is not known. For more in-depth discussion of considerations in the design of embedded training the reader is referred to Army Research Institute (1988); Witmer and Knerr (1991).

Personnel Performance and Training Research

The introduction of the helmet-mounted display and associated capabilities of the Land Warrior System will significantly alter how the soldier views scenes, interprets what is seen, makes decisions, and takes action. Although we have identified some of the performance training implications, many questions for research remain:

- What design considerations (processing speed, access rate, memory, display capabilities, communications, interfaces, menu structures) will be affected by training requirements?

- What design attributes will mitigate skill decay?
- How has the need for training evaluation and feedback affected the design?
- Should display characteristics be different for the trainer and the trainee? If so, how?
- How much training fidelity is required?
- How long will it take to gain proficiency?
- Which human attributes are correlated with successful use of the system?
- What are the trade-offs between soldier quality and training time?
- How frequently must training be conducted to sustain proficiency?
- Will the system help instill confidence and cohesion? If so, how?
- How will the lack of or diminished observation capability affect command and control within the squad?
- How will the system help the soldier visualize the battlefield?
- How do we know that there is greater battlefield awareness, and does it make a difference in performance at the squad level?
- What is it that this system provides that facilitates the development of creative responses to a fluid battlefield situation?

IMPLICATIONS FOR DISPLAY DESIGN

At issue in the panel's work is the compatibility of the proposed new technology with the capabilities and limitations of the target population. Operational requirements specify that "no qualitative or quantitative changes in personnel requirements will result from fielding the land warrior. No new military occupational specialties (MOSs) will be required for operators and maintainers." The Land Warrior System, however, will be providing amounts of information to the individual soldier that may be orders of magnitude greater than are now provided. We discuss below the implications for display design of force quality, physical attributes, soldier acceptance, and performance design.

Quality Issues

At the leader level, in the infantry there is a higher percentage of lower-level sergeants in category IIIB and IV (approximately 37 percent) and staff sergeants (approximately 41 percent) than in the overall Army force (U.S. Army, Natick RD&E Center, 1994). As a result of retention, there may be sergeants who may not have scored as high on the ASVAB as new recruits. As the propensity of young men to enlist goes down, the challenge to maintain quality will become even more difficult.

The implications of this situation are significant. Potentially, the greatest cognitive workload is on the fire team leaders and the squad leaders, who may be the least able to manage it. It is questionable as to whether the majority of the sergeants of tomorrow will possess the cognitive abilities and skills necessary to operate as envisioned and thereby capitalize on the new capabilities provided by the Land Warrior technology.

Approximately 33 percent of the infantry who will use the Land Warrior System will be in categories IIIB and IV. People in these categories can be expected to read at a 7th grade level (see Table 2-1), which is not indicative of speedy information processing (Bowman et al., 1986). Display contrast resolution that does not meet the international standard (ANSI/HFS 100-1988) will negatively affect reading speed. In addition, the display design must accommodate the hearing and vision limitations of current and future soldiers.

Historically, in introducing new technology, the Army has experienced unintended effects, such as increases in demands for soldier quality, increased school training time, the need for different skills, and lower levels of equipment readiness. The trade-offs associated with changing personnel selection criteria, quality distribution, and training all have long-term cost implications for the design of the Land Warrior System.

For example, the ORD does require that there be no increase to entry level requirements for maintenance MOSs 31V and 39E. The maintenance implications on the quality of the force might be offset by contractor provided maintenance. A fully integrated system will require careful Built-in Test/Built-in Test Equipment design to facilitate fault isolation and repair or replacement.

Anthropometric Issues

One of the important questions that arises is the degree to which the helmet-mounted display system will be adjustable to the individuals who will use it (see Gordon et al., 1988). Will display units be interchangeable among individuals, or are they to be tailored for individual soldiers? What are the implications of individual tailoring for security questions, such as the ability to block or lockout enemy access. The answers to these questions may dictate the approach taken in terms of anthropometric design.

Traditional anthropometry is a measurement and classification procedure that allows for the design of individual items to proceed around known physical characteristics of the population sample under consideration (i.e., each dimension of the human body). Such approaches try to provide the best compromise between statistical norms for a group and the unique attributes of individuals. Intelligent design (e.g., modularization) can allow for customization even if the basic configuration is developed on a group basis.

The basic considerations of anthropometry and biomechanics are the form and fit of the item under consideration. Some biomechanical factors include the requirement that any head-mounted device must consider the question of weight and prolonged use as they affect muscular fatigue. Likewise, the designers of hand-held devices also need to consider anthropometric factors, especially if data

TABLE 2-3 Head Dimension Measurements for Males (Inches)

Head Measure	1st percentile	5th percentile	95th percentile	99th percentile
Bitragion coronal arc	12.71	13.07	14.75	15.15
Bitragion crinion arc	11.84	12.11	13.63	13.99
Bizygomatic breadth	5.03	5.19	5.91	6.07
Head breadth	5.48	5.63	6.33	6.5
Head circumference	20.99	21.37	23.37	23.88
Head length	7.09	7.30	8.21	8.40
Interpupillary breadth	2.24	2.31	2.8	2.91

entry is to be provided through keyboard systems. In a military context, it is relevant to consider use of hand-held devices with gloves as an issue that requires careful anthropometric design. Again, in context, voice entry while wearing some form of mask or exclusion garment may prevent optimal use of critical systems. Thus, decisions concerning conditions of use raise any number of questions concerning the fit between the operator and the device.

The current target audience description does not cite head dimensions as critical. Given the importance of a comfortable fit to the soldier, the current design specification for 5th to 95th percentile males omits 10 percent of the population. Table 2-3 shows the difference between 5th to 95th percentile and the 1st to 99th percentile on a few dimensions of head measurement. The differences are so small that the needed adjustments should be easy to achieve. Because helmet stability is critical to maintaining a stable image, changes in the requirements seem called for. Given the history of the design of helmets and head-mounted displays for aviators, in which micro-anthropometrics was critical, this change is particularly significant.

Although the Land Warrior System is intended for the infantry, others may need it in order provide adequate support to them. It is expected that these personnel would come from the division's support units that are attached to the brigade or battalion task force (artillery, air defense, combat engineer, maintenance, medical, supply and transportation, etc.). These military occupational specialties and the personnel that fill them differ from the infantry population, not least because they include women. From a design fit perspective, this is an important issue. Other future users include the Army National Guard and the Army Reserve, other services, and allied or joint Forces. The anthropometric differences, if any, between these various groups should be assessed.

Potential Soldier Acceptance

Soldiers' acceptance of the Land Warrior System displays will be driven by their confidence in their ability to use the device effectively under adverse conditions, the devices's attributes, and their views of the need for it in light of everything else that they have to carry into battle. Interviews with the rangers who tested the prototype revealed that they valued three components of the system: squad radio, the thermal sight, and the global positioning system (in unobstructed areas). These components were valued because they provided needed capabilities that were previously not available. Other components were not considered as useful because they did not work reliably, were too heavy or uncomfortable, or did not improve performance. Despite the fact that certain problems clearly biased the test personnel, they were clear about what they thought added value. These data are primarily anecdotal, based on experiences in a preliminary, field demonstration.[2] The results of a series of interview following each phase of the demonstration can be found in Salter (1993).

A soldier's confidence in a new system is a function of his proficiency, which is in part determined by how much training he has received and the complexity of the system. These factors are interrelated. The more complex the system, the more training a soldier will need to gain and sustain proficiency. Sufficient training is often not provided, for reasons of limited resources. As a result, soldiers require a longer period of time to gain confidence in a system, if ever. Research by Marshall (1947) established that a large percentage of soldiers do not fire their weapons in combat. If a soldier is not proficient with the system, he will not risk exposure to enemy fire. The Land Warrior System offers a solution to this problem by letting a soldier fire accurately while not exposing himself to enemy fire. At the same time, the rangers reported that battlefield mobility was reduced by the night vision system. The complete and integrated Land Warrior System will be perceived to be complex.

When new systems are effective and well used, they can become a crutch without which the soldier or the unit may fail to respond. In the early days of TACFIRE (an artillery tactical fire control system), some artillery unit commanders found that their units would not respond to a call for fire when TACFIRE was inoperative. The personnel had become so dependent on the system that they had lost the ability to fire without it. Overreliance on technology can make a unit vulnerable if field-expedient training is not conducted.

Soldiers generally expect that a new system will be effective, reliable, and simple to operate, repair, and maintain; will reduce workload or improve effectiveness; and will fit comfortably. For the infantry soldier, if a new system does not help him shoot, move, communicate, or survive, it may be dropped along the march in order to lighten his load.

---

[2]Personal communication with soldiers at Fort Benning who participated in the SIPE field test.

Many leader tasks may be made easier with the Land Warrior System. For instance, reporting could be done photographically. Although this eases the workload associated with reporting, it may drastically increase the workload for those who receive and must interpret the messages. The new system will also require the mastering of some level of operator and maintenance tasks. The danger is that it creates a demand for information that requires the squad leader to change what he does and how he does it. The impact could affect the survivability or effectiveness of the squad when the leader's attention is not properly distributed.

Comfort is an essential ingredient to an infantry soldier who is carrying a heavy load; it usually involves form, fit, weight, and balance. Proper fit contributes to confidence. Conversely, a poor fit leads to frustration and anger. Soldier frustration with the initial prototype was reported to be very high (SIPE).

Performance Design Issues

Providing remedies to the problems of cognitive or information overload is not easy because of the various factors that affect work and the variations in emotional reaction to a variety of perceived risks, physical demands, and surprise. Technology is not the only driver of cognitive workload and may not be the principal one; communication and coordination tasks are major workload factors. Job tasks not related to a system impose significant workload. There is no doubt that workload will go up as a result of the Land Warrior System because new tasks will have to be performed that have not been performed previously. The question is one of penalties and payoffs.

The complex technology and all-pervasive impact of the proposed Land Warrior System on infantry soldiers raises a number of basic issues and design considerations. The proposed concept also raises doctrine questions about the autonomy of the individual soldier and about current individual skills and training requirements.

The requirement that no qualitative or quantitative changes in personnel requirements are to result from fielding the Land Warrior System (U.S. Department of the Army, 1994) will be an engineering design challenge. Contractors will have to translate this requirement into engineering criteria. For the helmet-mounted display, human factors engineers will want to know about user work requirements associated with field of view, field of regard, resolution, polarity, contrast, and brightness. They will also want to know about soldier ability requirements, such as spatial orientation, perceptual speed and accuracy, visual acuity, division of attention, and eye dominance. Design engineers will want to know system requirements, such as the required field of view, the mean time between failures, and how much accuracy is required. The software programmer will want to know what information is required and when, the required rate for updating data, the required refresh and update rate of the display, the definition (pixel by pixel) of symbols, and what control mechanisms accomplish what actions. Engineering psychologists will be concerned about degraded stimulus conditions, multiple input channels, pattern recognition, attention allocation and control, and individual differences. Answers to these questions are contingent on the tasks performed, the conditions of performance, the measures of effectiveness used, and the level of proficiency required (Zubal et al., 1990, 1993). Assumptions will be made to answer many of these questions. The questions warrant research in order to establish meaningful threshold values and conditions under which those values are valid.

CONCLUSIONS

Ultimately the success or failure of the Land Warrior System will depend on the individual soldier. There is a broad range of individual differences in the youth population and large variances around specific cognitive, physical, and psychomotor measures. Historically, personnel selection and soldier training have been used to mitigate this naturally occurring variance. Effective personnel selection requires knowledge of the human attributes that correlate with successful performance. For the Land Warrior System, little is known about the relationship between design attributes, human attributes, and successful performance.

Training time is limited. Embedded training solutions can assist if the hardware and software requirements needed are considered in the specifications for computer memory, processing speed, input and output devices, display characteristics, and system architecture. The current space, weight, and power constraints may restrict the effectiveness of an embedded training solution.

New equipment must fit properly and comfortably if soldiers are to use it willingly. Careful consideration needs to be given to ensuring that the design meets the anthropometric and biomechanical requirements of the user population. In the case of the helmet-mounted display, comfort and stability are dependent on design. The fit of the helmet may provide an unstable platform for the dismounted infantry soldier; this issue warrants attention. How to suppress the vibration in a helmet-mounted display that is caused by walking is also a question for research and design.

The amount of information that the Land Warrior System will provide to the individual soldier may be orders of magnitude greater than the information now provided. The profile of the target audience shows that a large number of potential infantry squad leaders are in the lower cognitive categories of military personnel and can be expected to read at a 7th grade level. Their ability to perform successfully needs to be closely evaluated. The increased cognitive burden is likely to be placed on the two least experienced leadership positions in the platoon—the squad leader and the platoon leader.

People who can work effectively with a rapid flow of information in a high-stress environment, in which decisions are made with less than full information, must be able to quickly determine the relevance and importance of the information sent or received. The needed attributes include adaptability, tenacity, the ability to learn quickly from experience and to work as a team, innovativeness and resourcefulness, cultural awareness, and tolerance. Some of the more interesting research questions in the area of personnel selection include:

- Which human characteristics (cognitive and non-cognitive, physical, psychomotor, etc.) are related to and predict success with Land Warrior?
- What tests and techniques can be used to select Land Warrior users based on temperament, values and attitudes?
- Will existing tests predict successful or superior performance?
- What impact will the LWS have on squad leader selection?

3

Situation Awareness

Situation awareness is a critical element of successful performance in the combat environment. The battlefield poses a variety of challenges to situation awareness: information overload, nonintegrated data, rapidly changing information, and a high degree of uncertainty brought on by lack of needed information. Overcoming these problems is a major goal of the Land Warrior System. Evaluating the degree to which proposed system designs actually provide benefits to situation awareness and help the soldier to think and act quickly, however, is a critical issue that needs to be addressed through careful design testing. Assessment of situation awareness in a systematic fashion will allow potentially critical problems to be detected, many of which are outlined in this chapter. The final designs selected must provide soldiers with the situation awareness they need to be successful in performing their many functions. Although it is beyond our current understanding to specify how much situation awareness is enough, researchers in the area believe that good performance is linked to good situation awareness (Endsley, 1995).

The chapter begins with a discussion of the individual characteristics that affect situation awareness, including perception, attention allocation, working memory, long-term memory stores, goal-directed behavior, and individual differences in relevant abilities. It then covers factors related to the task and the display system, including workload, complexity, automation, and environmental stressors. Next we consider the situation awareness of the combat unit as a team and how information is distributed, as well as how the display design can be expected to provide the specific type of support that soldiers need to perform their tasks. We then outline various approaches to measurement of situation awareness, including process indices, direct measures, behavioral measures, and performance measures. The chapter ends by summarizing the key points and key design recommendations in support of situation awareness.

THE SOLDIER'S SITUATION AWARENESS

A person's situation awareness can be described as his or her state of knowledge or mental model of the surrounding situation or the environment. It is not just spatial orientation but includes an understanding of the dynamics of the situation and the actions that are expected to take place in the future. Many definitions have been developed, some very closely tied to the aviation domain and some more general (see Dominguez, 1994, and Fracker, 1988, for reviews). A general, applicable definition describes situation awareness as "the perception of the elements in the environment within a volume of time and space, the comprehension of their meaning and the projection of their status in the near future" (Endsley, 1988a). It includes not only perceiving or attending to information, but also the integration of multiple pieces of information and determination of their relevances to one's goals, as well as the ability to forecast future situation dynamics providing for timely and effective decision making.

Situation awareness requirements for soldiers consist of the dynamic information needed to support each of their tasks and objectives. Although the specific objectives and tasks of the soldier vary from mission to mission, some common critical tasks can be hypothesized to include: detection and identification of targets, identification of terrain features, navigation and localization of self and others, engagement to include fire and maneuvering, communications between and within units, mission rehearsal planning and replanning, and development of tactics.

The elements of the environment that form the situation awareness requirements for the infantry soldier are as many as required to support each of these goals. They can include factors relevant to both the local and the global situation, as shown in Figure 3-1. Global situation awareness needs can be construed to include one's location within a broad geographical area, navigation information such as the relative location of important features, the current location and direction of movement of other units (friendly and enemy) and current commands and directions from headquarters. All these factors are relevant to the soldiers' ability to navigate and plan strategically to meet their goals. In order to know where to move to, and where not to, this type of information is critical.

Local information needs can include the location of a desired target in the immediate environment, the identity (friend, foe or neutral) of an entity under current targeting, terrain and object location (as needed for basic mobility and maneuvering), and cueing of the presence and movement of enemies in the immediate environment. This information is critical to the soldier's basic aware-

Global situation awareness	Local situation awareness
location of self	target identification
location/movement of other units	target location
commands/directions from headquarters	terrain/object distance
navigation information	cueing of hostile presence

FIGURE 3-1 Situation awareness needs of the infantry soldier.

ness of hostile presences around him and the ability to act and react quickly in accordance with mission goals.

Both global and local needs for situation awareness are critical for effective functioning in a given environment. The local information is required for effectively acting to meet immediate needs. The global information is required for employing oneself effectively in concert with other units to meet strategic goals.

It should be noted that the advent of helmet-mounted displays in this domain may affect each of the two types of situation awareness needs quite differently. For example, although the displays being considered may produce better situation awareness at the global level, it may also reduce local situation awareness by removing the soldier's attention from the immediate surroundings. (This issue is discussed in more detail later). This examples illustrates two major issues to be noted: (1) there is a definite need for the soldier to maintain an adequate level of awareness across all elements of the environment, and (2) due to limits on attention, gains in situation awareness on some elements can occur at the expense of losses on other elements. The potential for these trade-offs to occur needs to be very closely examined during the design process in order to ensure that adequate situation awareness is provided across all requirements.

FACTORS AFFECTING SITUATION AWARENESS

Endsley (1988a, 1994, 1995) has proposed a framework model of situation awareness based on information-processing theory (Wickens, 1992), Figure 3-2. The model includes a number of factors that influence situation awareness and that can be seen to be potentially affected by the proposed helmet-mounted display.

Individual Factors

It is believed that situation awareness is primarily restricted by the limits of attention and the capacity of working memory, which constrain a person's ability

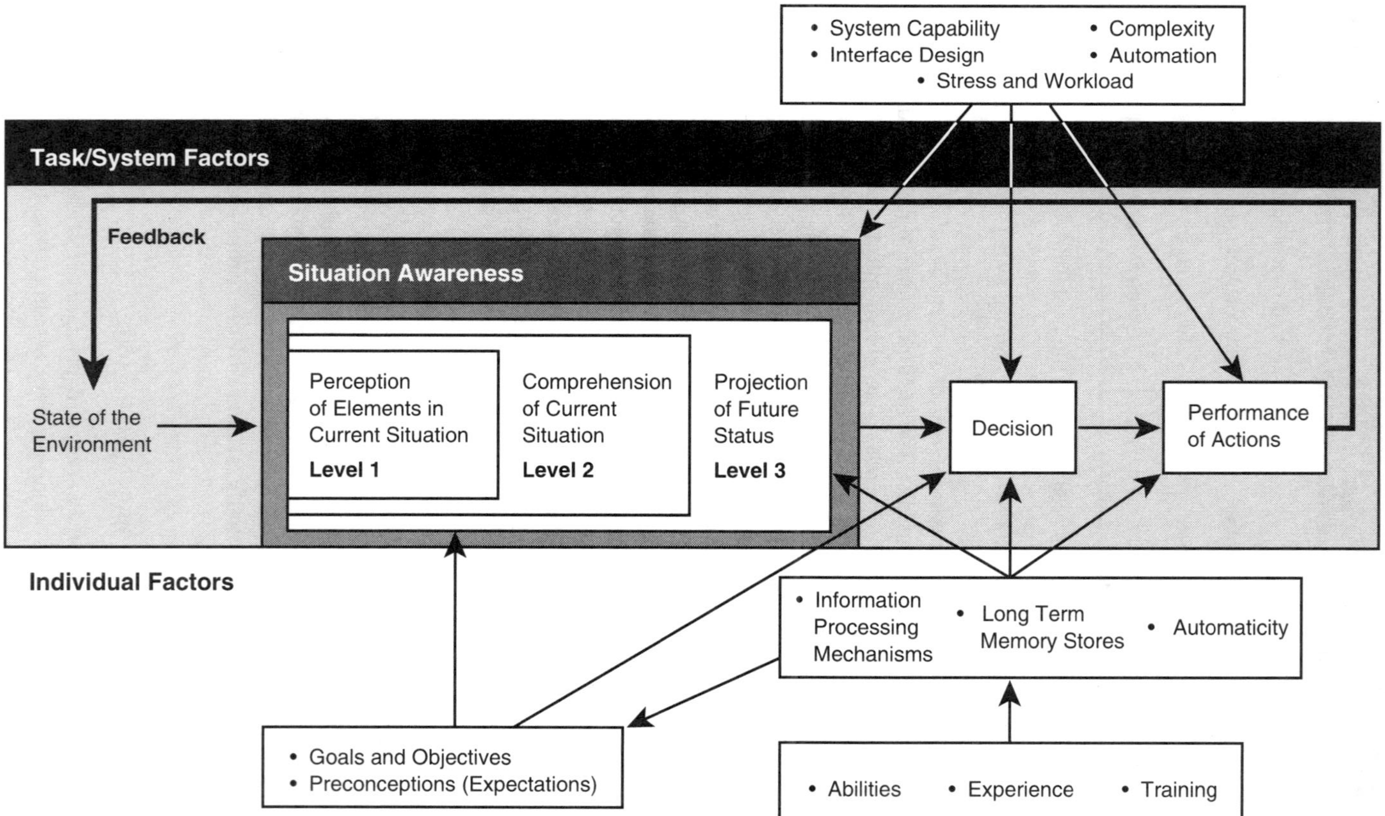

FIGURE 3-2 Model of factors affecting situation awareness. Source: Endsley (1995).

to take in and process multiple channels of information. Long-term memory stores, when they have been developed through experience, most likely in the form of schemata and mental models, are proposed to circumvent these limits by providing for the integration and comprehension of information and the projection of future events (higher levels of situation awareness), thus significantly off-loading working memory and efficiently directing perceptual processes. The use of mental models in achieving situation awareness is considered to be dependent on the individual's ability to match critical cues in the environment and elements in the mental model. Schemata of prototypical situations are incorporated in this process and in many instances may also be associated with scripts to produce single-step retrieval of actions from memory, thus providing for very rapid decision making, as has been noted by Klein (1989).

In addition, information held in working memory (current goals, expectations, other situational information) is believed to affect attention deployment and the perceptual process. Situation awareness is largely influenced by a person's goals and expectations, which influence how attention is directed, how information is perceived, and how it is interpreted. The model shows this goal-directed, top-down processing operating in tandem with bottom-up processing, in which highly salient cues activate appropriate goals and mental models. Thus, situation awareness is the result of an ongoing process of alternating between goal-directed processing and data-driven processing based on a theorized link between goals and mental models. In the sections that follow, we discuss several factors that affect the accuracy and completeness of that situation awareness that soldiers derive from their environment.

Perception

First, basic perceptual processes may be affected. Trade-offs may occur between the degree to which the helmet-mounted display enhances perception (with night vision, for example) and the degree to which it interferes with normal perceptual processes, including hearing and vision. Situation awareness can be negatively impacted if the device prevents perception of important environmental information or creates misperceptions.

Attention Allocation

The way in which attention is employed in a complex environment with multiple competing cues is essential to determining which aspects of the situation will be processed. A major factor of situation awareness that may be negatively influenced by the helmet-mounted displays is the soldier's attention allocation. Attentional narrowing has been cited as a major factor in errors related to situation awareness (Endsley, 1995). The display creates a new source of information that may compete with the outside world for the soldier's attention, rendering him susceptible to attentional narrowing (focusing narrowly) on the display, thereby missing important information in the external environment. The tendency to focus attention on electronic information displays is encouraged by their high degree of perceptual salience. With a display located directly in front of the soldier's eye, this hazard may become even more pronounced. Alternative display technologies that do not directly interfere with the intake of environmental information may need to be considered in order to minimize this problem; however, it is doubtful that the danger can be completely eliminated. It may also be necessary to consider uses (e.g., distributing the capability among team members) that compensate for potential losses of local situation awareness when using the system.

Working Memory

Once taken in, information must be integrated with other information, compared with goal states, and projected into the future—all activities that make heavy demands on working memory. The limits of a soldier's working memory should be considered, particularly since working memory can be further reduced under stress, such as that of combat (Hockey, 1986; Mandler, 1979). The design of any electronic device that is introduced into this type of environment must take the limits of working memory into account with an interface that minimizes requirements to memorize commands, syntax, or other information. Its presentation in an easy to process format will also be critical for minimizing demands on the soldier's limited processing resources.

In this context, it is critical that the information provided truly support the attainment of the soldier's goals and not impose extra demands or provide extraneous information. New information cannot be provided to the soldier without some costs; there is a danger of going from too little information to too much information. The presence of a new source of information that must be integrated with information in the environment can degrade situation awareness by imposing extra processing requirements. Due to the limited ability to perceive and process information, significant difficulties may be encountered unless stringent measures are taken to integrate multiple sources of information, reduce extraneous information, simplify the format of information presentation, and integrate the presentation of information with the soldier's tasks.

Long-Term Memory Stores

Long-term memory stores in the form of mental models or schema are hypothesized to play a major role in dealing with the limitations of attention and working memory. With experience, people develop internal models of the system they operate and the environments they operate in. These mental models serve to help direct limited attention in efficient ways, provide a means of integrating information without loading working memory, and provide a mechanism for generating projections of future system states. The new technologies being considered for the soldier may promote good situation awareness by allowing information to be presented in a format that is more compatible with these mental models. The ability to provide concrete, up-to-date maps of the environment with explicit representations of tactical units can reduce mental processing requirements. For instance, egocentric displays have been found to be superior to exocentric displays for self-locomotion by improving the compatibility of the direction of motion in the display and the motion in the environment (Wickens et al., 1989). It may also be more advantageous to receive certain information visually instead of audibly, or vice versa.

Goal-Directed Behavior

Goals are also important for situation awareness. Human information processing is seen essentially as alternating between data-driven (bottom-up) and goal-driven (top-down) processing. In goal-driven processing, attention is directed across the environment in accordance with active goals. A person actively seeks information needed for goal attainment, and the goals simultaneously act as a filter in interpreting the information that is perceived. In data-driven processing, perceived environmental cues may indicate new goals that need to be active. Dynamic switching between these two processing modes is important for successful performance.

A significant problem can occur if people are not receptive to the perception of relevant data (e.g., slight sounds indicating an enemy sneaking up) while engaged in essentially goal-directed behavior (e.g., searching for information or providing information across a network). A display that significantly loads or interferes with the broad-based perception of information from the environment will therefore be of particular concern. Good display design will enhance the salience of critical environmental cues and will make the attainment of information relevant to a particular goal as easy as possible.

Individual Differences

Because anecdotal evidence suggests that some people are better at maintaining a high level of situation awareness, it may be important to consider factors that relate to individual differences in this area during the training and selection process. On the basis of the model presented here, several factors have been hypothesized to be important determinants of variability among individuals in terms of situation awareness capability: (1) spatial abilities, the degree to which one can mentally visualize and manipulate objects and also visualize one's own orientation relative to those objects, (2) attention abilities, specifically attention sharing as needed to achieve situation awareness in a complex environment,

(3) memory, including working memory capacity and the quality and quantity of long-term memory stores, (4) perception, the ability to rapidly perceive and assimilate new information, and (5) logical and analytical skills that may be useful in searching out information and piecing it together (Endsley and Bolstad, 1994).

Testing these hypothesized factors against a measure of situation awareness, Endsley and Bolstad (1994) found that, among pilots, those with higher situation awareness had better spatial abilities and perceptual speed. They also found partial support for a link between higher situation awareness and better pattern-matching abilities and attention-sharing abilities. The degree to which these findings can be generalized to an infantry population is not known. This list may provide a starting point for generating research on factors that are relevant to situation awareness for the soldier, however.

Endsley and Bolstad (1994) found a 1 to 10 ratio in situation awareness capability separating the pilots in their study. Scores were furthermore found to be fairly stable for an individual tested at different times, indicating that some people may indeed be better at maintaining situation awareness (either due to innate ability, different strategies employed, or differences in training and experience). This issue needs further exploration in the infantry population. It may be possible to select individuals with better situation awareness capabilities for tasks that involve the new proposed technologies, or to better train them to avoid problems and use effective situation awareness skills, if such skills can be identified.

Task and System Factors

In addition to individual factors, many features of the environment may affect the soldier's awareness. The task and system factors discussed in this section need to be considered when designing an information support system.

Workload

The link between situation awareness and workload is depicted in Figure 3-3 (Endsley, 1993). With low to moderate workload, the level of situation awareness a person has can be independent of workload level. One may have low situation awareness and may not be working very hard to achieve a higher level. Or one may have high situation awareness without having to work very hard (through the benefits of a well-designed system). One may be working fairly hard and may be rewarded with a high level of situation awareness, or one may still have low situation awareness, if one's efforts are ineffective or one misinterprets the information acquired.

At very high levels of workload, situation awareness may suffer, however. If the volume of information and number of tasks are too great, only a subset of

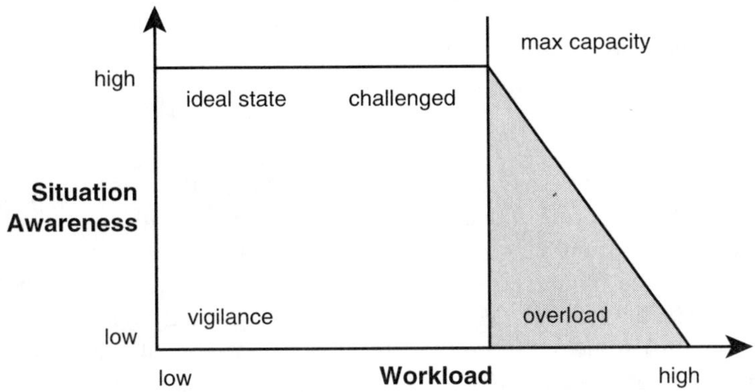

FIGURE 3-3 Relationship between situation awareness and workload. Source: Endsley (1993).

information can be attended to. Or one may be actively working to achieve situation awareness, yet suffer from erroneous or incomplete perception and integration of information. A display that overloads the soldier can lead to low levels of situation awareness. It is important that the technologies implemented in the battlefield do not further increase workload, particularly during high-workload tasks.

Poor situation awareness can also occur under low workload, however, in which case the operator may have little idea of what is going on and not be actively working to find out due to inattentiveness, vigilance problems, or low motivation. Although electronic information systems represent a way of productively increasing workload (and information) during periods of low workload, it is critical that this not occur at the expense of maintaining vigilance regarding the immediate environment. Further workload considerations are addressed in Chapter 6.

Complexity

A major challenge to maintaining good situation awareness is the complexity of many of the systems that must be operated. The more complex the systems are to operate, the greater the increase in the mental workload required to achieve a given level of situation awareness. When that demand exceeds human capabilities, situation awareness will suffer. System complexity may be somewhat moderated by the degree to which the person has a well-developed internal representation of the system to aid in directing attention, integrating data, and developing the higher levels of situation awareness—mechanisms that may be effective for coping with complexity. Developing those internal models, however, requires a considerable amount of training and may be beyond the capabilities of many soldiers, as indicated by their ASVAB scores, as we discussed in Chapter 2.

Automation

Situation awareness may be negatively affected by the automation of tasks. System operators working with automation have been found to have a diminished ability to detect system errors and subsequently perform tasks manually in the face of automation failures, compared with entirely manual performance on the same tasks (Billings, 1991; Moray, 1986; Wickens, 1992; Wiener and Curry, 1980). Although some of this problem may be due to a loss of manual skills with automation, situation awareness is also a critical component (Endsley and Kiris, 1995).

Operators who have lost their situation awareness may be slower to detect problems and also may require extra time to reorient themselves to relevant system parameters in order to proceed with problem diagnosis and the resumption of manual performance when automation fails. This has been hypothesized to occur for a number of reasons, including (a) a loss of vigilance and increase in complacency associated with the assumption of a monitoring role under automation, (b) the difference between being an active processor of information in manual processing and a passive recipient of information under automation, and (c) a loss of or change in the type of feedback provided to operators concerning the state of the system under automation (Endsley and Kiris, 1995). The degree to which the automation of tasks is incorporated in the helmet-mounted display needs to be carefully examined for this potential impact.

Stressors

Several types of stressors in the combat environment may affect situation awareness, including physical stressors—noise, vibration, heat or cold, lighting, atmospheric conditions, boredom or fatigue, cyclical changes—and social/psychological stressors—fear or anxiety, uncertainty, the importance or consequences of events, self-esteem, career advancement, mental load, and time pressure (Hockey, 1986; Sharit and Salvendy, 1982). A certain amount of stress may actually improve performance by increasing attention to important aspects of the situation. A higher amount of stress can have extremely negative consequences, however, as accompanying increases in autonomic functioning and aspects of the stressors can act to demand a portion of a person's limited attentional capacity (Hockey, 1986).

Stressors can affect situation awareness in a number of different ways, including narrowing attention. With perceived danger, a decrease in attention has been observed for peripheral information—those aspects that attract less attentional focus (Bacon, 1974; Weltman et al., 1971). Broadbent (1971) found an increased tendency to sample dominant or probable sources of information under stress. This is a critical problem for situation awareness, leading to the neglect of certain elements in favor of others. In many cases, such as in emergency conditions, factors outside the person's perceived central task are the ones that prove to be lethal.

Premature closure, arriving at a decision without exploring all the information available, has also been found to be more likely under stress (Janis, 1982; Keinan, 1987; Keinan and Friedland, 1987). This includes both considering less information and attending more closely to negative information (Janis, 1982; Wright, 1974). Several authors have found that the scanning of stimuli under stress is scattered and poorly organized (Keinan, 1987; Keinan and Friedland, 1987; Wachtel, 1967).

Another way in which stress may impact situation awareness is through decrements in working memory capacity and retrieval (Hockey, 1986; Mandler, 1979). The degree to which working memory decrements affect situation awareness depends on the resources available to the individual operator. In tasks for which achieving situation awareness involves a high load for working memory, a significant impact on situation awareness levels 2 and 3 would also be expected. If long-term memory stores are available to support situation awareness, less effect is expected.

Although anxiety is a common stressor in the battlefield, other stressors, such as fatigue and environmental conditions (cold, heat, humidity), can also take a significant toll on performance and situation awareness through these same mechanisms. To a certain degree, the impact of stressors on situation awareness is a given part of the combat environment. Many new proposed technologies can exacerbate these effects, however, if they interfere with scanning of relevant information in the environment, load working memory, or encourage dependence on highly perceptually salient technological information sources. They can also be designed to mitigate these potential problems by providing an easy to access overview of critical information that might otherwise be neglected or lost from working memory under stress. The effects of stressors on soldier performance are discussed further in Chapter 6.

INFORMATION DISTRIBUTION

Within the combat unit, individuals must work together as a team to carry out actions effectively, so overall team situation awareness becomes important. In this context, each team member has a specific set of elements of situation awareness about which he is concerned, as determined by his responsibilities within the team. There is some overlap of each team member's situation awareness requirements, and this subset of information is the basis for much of team coordination. Coordination may occur as a verbal exchange or may be provided through the system display or by some other means (e.g., nonverbal communication). The quality of the team members' awareness of shared elements (as a state of knowledge) may serve as an index of team coordination or human-machine interface effectiveness. Thus, an important issue is the degree to which the helmet-mounted display will affect situation awareness across the team. It is possible that it may be improved for one individual, but not for others, if the display does not support needed information transfer across the team, or if it physically interferes with other means of information transfer (such as direct verbal and nonverbal exchanges).

DESIGN DRIVERS

The design of information displays should be informed by a careful consideration of the type of support the soldier really needs in achieving better situation awareness and what factors currently act to limit it in his environment. In this light, the real utility of the proposed technologies may be in providing: (1) sensory enhancement, improving the soldier's ability to localize targets and self and to navigate in the environment, (2) more dynamic information, keeping the soldier and commander up to date on changes and situational factors in the field, (3) information sharing between team members, supporting planning and dynamic decision making, (4) distributed decision making providing information across teams and between headquarters and teams, and (5) strategic decision making, allowing soldiers to look at information in different ways, thus supporting different integration, comprehension, and projection possibilities.

An analysis of the impact of the different proposed information technologies shows that they may be expected to impact situation awareness in different ways across local and global needs (see Table 3-1). On one hand, for example, GPS may be expected to dramatically improve global situation awareness. The improved information flow from headquarters to the soldier (and back) can be expected to improve the situation awareness of all parties in their knowledge of the global picture (e.g., the latest location, movements of friendly and hostile units). In addition, the graphics capabilities afforded by the helmet-mounted display should provide an improvement in situation awareness over current piecemeal audio technologies. On the other hand, the helmet-mounted displays and night vision system may reduce local situation awareness by drawing the soldier's attention away from the immediate environment and into the virtual one or by inducing certain misperceptions regarding the actual location (distance) of objects.

MEASUREMENT

As many factors surrounding the helmet-mounted displays and other technologies may act to both enhance and degrade situation awareness in the environment of the infantry soldier, significant care should be taken in evaluating the

TABLE 3-1 Impact of New Technologies on Situation Awareness

Potential New Technologies	Global Situation Awareness	Local Situation Awareness
	Location of self	Target identification
	Location/movement of other units	Target location
	Commands/directions from headquarters	Terrain/object distance
	Navigation information	Cueing of hostile presence
Helmet-mounted display	Improved quality/quantity of information	Decreased situation awareness of immediate environment
Night vision goggles	No impact	Improved situation awareness of environment in low light, possible misperceptions
Thermal weapon sight	No impact	Improved localization/ identification of targets
Laser range finder	No impact	Improved localization/ identification of targets
Audio	Improved transfer of information between team members	Improved transfer of information between team members

impact of proposed concepts on situation awareness. Only by testing new design concepts in carefully controlled studies can their actual impact be identified. This testing should include an examination not only of how the information technologies affect basic human processes such as the accuracy of perception, but also of how they affect the soldiers' situation awareness (across all elements) when used in dynamic and complex field settings in which multiple sources of information compete for attention and must be selected, processed, and integrated in light of dynamic goal changes. Real-time simulations employing the helmet-mounted display can be used to assess the impact of the system by carefully measuring soldier performance, workload, and situation awareness. Direct measurement of situation awareness during design testing is recommended to provide sufficient insight into potential costs and benefits of design concepts for soldiers' situation awareness, allowing a determination of the degree to which the design successfully addresses the issues we have discussed.

Situation awareness is a relatively new focus in system design. Like other constructs, such as workload, a variety of measures have been used in its assessment. At this time it is not possible to identify a particular measure as the "gold standard." The following review provides the pros and cons of several measures. Each of these may be useful at different times in the design process.

High-level performance measures of combat (e.g., kills and losses), as collected under the constrained conditions of simulation testing, are often not sufficiently granular or diagnostic of differences in system designs. Whereas one design concept may be superior to another in providing the soldier with needed information in a format that is easier to assimilate with his needs, the benefits may go unnoticed under the constrained conditions of simulation testing or due to extra effort on the part of soldiers to compensate for a design's deficiencies. If situation awareness is measured directly, it will be possible to select concepts that promote it increasing the probability that soldiers will make effective decisions and avoid poor ones. Problems with situation awareness, frequently brought on by data overload, nonintegrated data, automation, complex systems that are poorly understood, excess attention demands, and many other factors can be detected early in the design process and corrective changes made to improve the design.

Multiple types of testing are desirable. At the most simple level, the system needs to be tested in part-task studies under well-controlled laboratory conditions. This type of testing examines the degree to which certain design features affect human performance in conducting very explicit tasks—for example, time and error rates for finding required information or entering information with different display formats. This testing needs to be very carefully controlled in order to detect potential problems with perceptual tasks (finding information, accurately perceiving the information, detecting information), motor tasks (entering information, range of motion tests, physical interference with environment), and cognitive tasks (finding needed displays, making decisions). This type of testing facilitates the selection of design features that enhance performance.

For situation awareness, an even more critical type of testing involves simulations of the task environment at medium to high levels of fidelity. Scenarios developed for this type of test should incorporate more realistic environmental task loads (e.g., conducting multiple tasks within a realistic mission scenario). The tests should include both expected and unexpected events and factors—for example, enemies that are not where they were projected to be, sneak attacks, loss of friendly forces, replanning from headquarters. These tests should include multiple types of information coming from multiple sources. The objective is to provide a testing environment that accurately depicts these features of the operating environment, so that the utility of the devices can really be explored. Because situation awareness in particular is very affected by attentional deployment and competing demands, an accurate picture of the impact of any new technology can be examined only by incorporating these issues, showing how the technology may affect situation awareness (across all of its elements) when used in demanding conditions. This type of design testing is essential for the task of designing an integrated human-centered system.

Finally, testing under field scenarios should be considered. This provides the highest level of realism, but also the lowest level of control and measurability of the issues of interest. Although situation awareness may be difficult to assess directly under these conditions, it may be inferred from operational performance.

Scenarios for field testing usually include realistic mission scenarios and environmental conditions (e.g., sweat occluding vision, equipment that shifts in use).

The process model in Figure 3-4 represents the issues involved in selecting measures of situation awareness. This model shows the stages involved in the sequence from perception to action. Although they are shown as separate stages for simplicity in narration, it should be noted that these stages may be very closely coupled. Moderating factors that may influence each stage are shown on the left. On the right, classes of measures appropriate to each stage are shown. Measures at each stage are discussed in the next sections including the advantages and disadvantages of each (Endsley, 1996).

Process Indices

An examination of the assessment processes people use to acquire situation awareness may provide information about how soldiers allocate their attention in using a particular system design. It may indicate the relative priority of different types of information or the relative utility of information sources. In general, however, process measures provide only an indirect indication of operator situation awareness. Eye-trackers may indicate how attention is used the process of acquiring situation awareness, typical scan patterns, and relations between elements. Studying the verbal communications between soldiers may also suggest the types of information that are missing from displays, verbal techniques used for acquiring situation awareness, and differences in situation awareness strategies among individuals.

Verbal protocols may provide some useful information on not only what is attended to, but also how that information is integrated and used. Significant difficulties in processing and using the data provided by verbal protocols must be dealt with by the experimenter, however, if this technique is to be used successfully.

Each of these techniques can be viewed as providing useful partial information on processes of acquiring situation awareness, from which some inferences may be possible. Because verbal communications and verbal protocols take place in a very limited time frame, however, they cannot be regarded as complete representations of what people attend to or process. Eye-trackers and information acquisition methods do not provide any insight on how the information is used or combined to form higher-level situation awareness.

Direct Measures

Two types of measures have been developed for assessing situation awareness directly: subjective techniques and questionnaires.

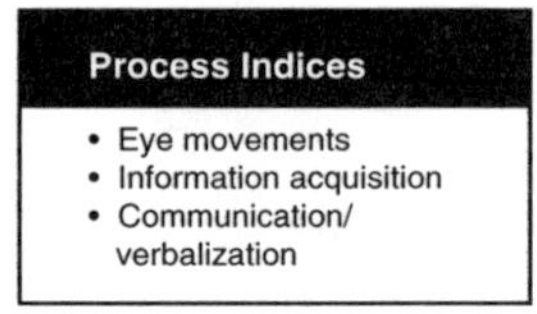

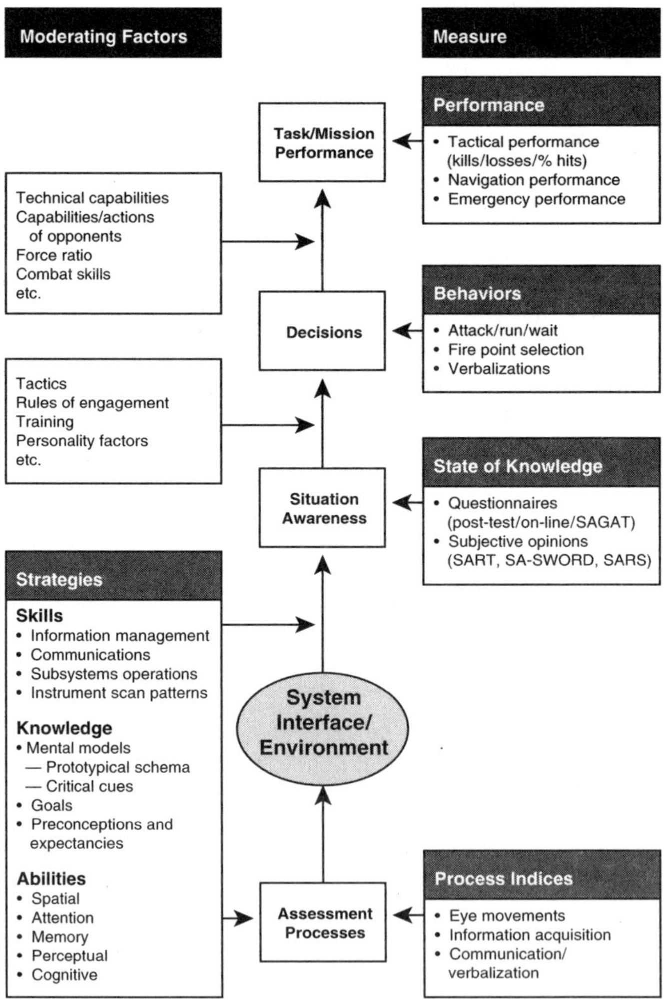

FIGURE 3-4 Process model of situation awareness measures. Source: Adapted from Endsley (1995a).

Subjective Technique

Subjective estimation of situation awareness can be made by individual soldiers or by experienced observers. Subjective assessment is very attractive in that it is fairly inexpensive and easy to administer. In addition to allowing evaluation of design concepts in simulation studies, subjective techniques can be easily applied in less controlled, real-world settings. Certain limitations, however, constrain the interpretation of subjective evaluations of situation awareness.

Self-ratings usually involve a subjective estimation of how much situation awareness a particular person feels he or she has when using a given system design. Self-ratings may not necessarily provide an accurate quantification of situation awareness, however, because people may not know about their own inaccuracies or what information they are unaware of and have a limited basis for making such judgments. In addition, self-ratings may be highly influenced by self-assessments of performance, thus becoming biased by issues that are beyond the construct of situation awareness. These self-ratings may be useful, however, as an assessment of the soldier's degree of confidence in his situation awareness (which can also affect decision making).

One of the best-known subjective scales is the situational awareness rating technique (SART) developed by Taylor (1990). SART has individuals rate a system design based on the demand on attentional resources, the supply of attentional resources, and understanding of the situation provided. As such, it considers individuals' perceived workload (supply and demand on attentional resources) in addition to their perceived understanding of the situation. SART has been developed on the basis of items that air crew report to be important to situation awareness and has been shown to be sensitive to workload variations.

In another approach to developing a standardized subjective measure of situation awareness, Vidulich and Hughes (1995) used a modified version of the subjective workload dominance (SWORD) technique to obtain subjective evaluations of the situation awareness provided by displays. SA-SWORD has subjects provide a comparative preference for displays on a nine-point scale, on the basis of their beliefs about the amount of situation awareness provided by each. The technique has not been validated for measurement of situation awareness, however.

Situation awareness may be assessed by subjective ratings of outside observers. An advantage is that trained observers may have more information than the subject about what is really happening in a given simulation, so their knowledge of reality may be more complete. A shortcoming is that observers have only limited knowledge about the subject's concept of the situation. Operator actions and verbalizations may provide useful diagnostic information on explicit problems (misperceptions or lack of knowledge) and an indication that certain information is known, supporting observer judgments. Actions and verbalizations cannot be taken to provide a complete representation of an operator's situation awareness, however. They may know many things that they do not mention or make an immediate response to as they are performing other tasks, for example. Observer ratings therefore provide only a partial indicant of a subject's situation awareness. Efforts to elicit more information (by asking questions or providing artificial tasks) may augment natural verbalizations, but this may alter the subject's distribution of attention, thus altering situation awareness.

Questionnaires

Questionnaires allow for the collection of detailed information about the subject's perceptions that can be evaluated against reality, thus providing an objective assessment of situation awareness on a detailed level. This type of assessment provides a direct measure and does not require subjects or observers to make judgments about situational knowledge on the basis of incomplete information, as subjective assessments do. This type of information can be gathered in one of three ways: post-test, during simulations, or during interruptions in the simulation.

A detailed questionnaire can be administered after the completion of each simulated trial, allowing ample time for subjects to respond to a lengthy and detailed list of questions. Memories of dynamic situation awareness will be less reliable with time, however; people have been shown to overrationalize and overgeneralize about past mental events (Nisbett and Wilson, 1977). Early misperceptions may be quickly forgotten as the situation unfolds over time. Post-test questionnaires will reliably capture situation awareness only at the very end of a trial. Kibbe (1988) used this technique to evaluate situation awareness as affected by automation of a threat recognition task. She found a retrospective recall measure to be insensitive to the automation and problematic.

One way of overcoming this deficiency is to ask subjects about their situation awareness while they are carrying out their simulated tasks. This strategy may alter situation awareness and task performance, however, as it can be regarded as providing an additional secondary task and intrusive.

To overcome the limitations of reporting on situation awareness after the fact, several researchers have used a technique wherein the simulation is frozen at randomly selected times, the system displays are blanked, and the simulation is suspended while subjects quickly answer questions about their current perceptions of the situation. Subject perceptions are then compared with the real situation based on simulation computer databases to provide an objective measure of situation awareness. The collection of data in this manner provides an objective, unbiased assessment of situation awareness that overcomes the problems incurred when collecting data after the fact, yet minimizes biasing due to secondary task loading or artificially cueing the subject's attention. The primary disadvantage of this technique involves the temporary halt in the simulation.

The situation awareness global assessment technique (SAGAT) is a global tool developed to assess situation awareness across all of its elements on the basis of a comprehensive assessment of operator requirements (Endsley, 1987, 1988b, 1990b). As a global measure, SAGAT includes queries about all operator situation awareness requirements, including Level 1 (perception of data), Level 2 (comprehension of meaning), and Level 3 (projection of the near future) components. It includes a consideration of system functioning and status as well as relevant features of the external environment. The approach minimizes possible biasing of attention, because subjects cannot prepare for the queries in advance (since they could be queried over almost every aspect of the situation to which they would normally attend).

SAGAT has been shown to have predictive validity, with SAGAT scores indicative of pilot performance in a combat simulation (Endsley, 1990a). Content validity was also established, showing the queries used to be relevant to situation awareness in a fighter aircraft domain (Endsley, 1990b). Empirical validity has been demonstrated through several studies that have shown that a temporary freeze in the simulation to collect SAGAT data did not affect performance and that such data could be collected for up to 5 or 6 minutes during a freeze without running into memory decay problems (Endsley, 1990b, 1995). A certain degree of measurement reliability has been demonstrated in a study that found high reliability of SAGAT scores for four individuals who participated in two sets of simulation trials (Endsley and Bolstad, 1994).

To apply this technique to the infantry combat task, a simulation environment in which soldiers perform realistic tasks with and without the aid of the proposed technologies is needed. (Applying the technique in real-world settings may be difficult or prohibitive.) A SAGAT battery of questions would also need to be constructed based on an analysis of the soldier's situation awareness requirements. This allows for domain-appropriate assessments of situation awareness and provides information on how the soldier's situation awareness is affected by each new technology.

Behavior Measures

Operators can be expected to act in certain ways on the basis of their situation awareness. Some information about situation awareness may, therefore, be determined from examining behavior on specific subtasks that are of interest. Behavioral indices could include time to make a response (verbal or nonverbal) and an estimation of correct or incorrect situation awareness as identified from verbalizations and appropriateness of a given behavior for a particular situation. Assessments of situation awareness based on these types of behavioral measures need to be viewed with caution, since they assume what the appropriate behavior will be, given situation awareness or lack of it. The assumptions may not necessarily be warranted. For example, a subject may choose not to immediately verbalize or respond to a given event or may employ different response strategies, thus confounding this type of measure.

Performance Measures

In general, performance measures provide the advantage of being objective and are usually nonintrusive. Computers for conducting simulation testing can be programmed to record specified performance data automatically, making them relatively easy to collect. Several limitations exist in using performance data to infer situation awareness, however. Global measures of performance (e.g. success in meeting a goal, kills and losses in a battle) are important as measures of situation awareness, however, they are limited. Because many moderating factors can influence the link between situation awareness and performance, global performance measures provide only an indirect indication of situation awareness.

Some definite task measures may readily present themselves for evaluating certain kinds of systems, but for others determining appropriate measures may be more difficult. An expert system, for example, may influence many factors in a global, not readily predictable manner. The major limitation of this approach stems from the interactive nature of situation awareness subcomponents. A new system to provide situation awareness on one factor may simultaneously reduce it on another, unmeasured, factor. In addition, it is quite easy for subjects to bias their attention to a single issue that is under evaluation in a particular study if they figure out the purpose of the study. Overall, relying exclusively on measures of performance on specific parameters can yield misleading results and should be viewed within the context of other types of measures.

CONCLUSIONS

A major purpose of the Land Warrior System is to improve the infantry soldier's situation awareness, which in turn is anticipated to improve his performance. There is some evidence that mission performance correlates with situation awareness in aircraft simulators. The link between performance of the infantry soldier and his level of situation awareness has yet to be studied systematically. Based on our review of the literature in this area, we draw the following conclusions regarding the helmet-mounted display and its potential role in situation awareness:

1. The helmet-mounted display system has the potential to enhance situation awareness by providing timely, more up-to-date information, and better sharing of information across team members, units, and geographic areas.
2. The helmet-mounted display may improve the soldier's situation awareness about global information (location of self and others in environment, communications with headquarters, navigation).

3. The helmet-mounted display may compromise the soldier's local situation awareness (location, presence of enemies, terrain and object perception) by competing for limited attention resources, affecting perceptual processes, or both.

4. Hand-held or wrist-mounted displays should be seriously considered as an alternative to the helmet-mounted displays in order to reduce the likelihood of negatively affecting the soldier's local situation awareness.

5. The system can reduce situation awareness if it poses a significant demand on mental resources or shifts the task load away from regular duties to system operation. These problems will be worse under stress and with higher levels of system complexity.

7. Significant increases in requirements for skills and abilities may be created by the helmet-mounted display, indicating the potential for changes in selection and training requirements to allow for acceptable levels of situation awareness and performance.

DESIGN GUIDELINES

The following design recommendations are based on our discussion of the cognitive mechanisms involved in achieving situation awareness.

1. The design of the display system should minimize the degree to which it is a physical barrier to acquiring environmental information (e.g., occludes or alters normal hearing and vision). It should enhance sensory input only when needed (e.g., targeting support, night vision).

2. The display design should minimize the degree to which it distracts attention (e.g., make the system removable, place it out of the normal line of sight).

3. The display design should minimize the cognitive load it places on the user by:

- providing integrated information (e.g., fusing information from different sources),
- providing easy user input of information (e.g., menus),
- minimizing memory requirements,
- reducing extraneous information,
- simplifying the format of information presentation,
- minimizing tasks,
- presenting information in a task-oriented sequence and grouping, and
- proving information in the needed format (e.g., egocentric maps).

4. The display should be designed to enhance situation awareness by providing salient cueing, directing attention to the most important information.

5. The display design should minimize complexity and avoid high levels of automation.

6. The system should provide new capabilities needed by the soldier, such as integrating information (as needed for decision making), comparing information to pertinent goal states, allowing a projection of future states, and providing support for human memory.

7. The display design should allow for easy sharing of information between team members and between the field and headquarters.

4

Visual and Psychomotor Factors in Display Design

In a working environment that is arguably the most dangerous of any current profession, image intensification and thermal imaging have extended the normal perceptual capabilities of the soldier and allowed vision to operate in conditions in which the unaided eye would be ineffective. The proposed helmet-mounted display is designed to allow various information sources to be displayed in front of one eye on a single screen, thereby reducing the time required to switch from one source to another.

A great deal is known about the human visual system and its strengths and limitations in a variety of conditions. The scientific evidence regarding visual and psychomotor factors is among the most critical the panel has assessed. In this chapter, we identify several human factors issues that should be carefully considered by the system's designers.

In our examination of visual and psychomotor attributes of helmet-mounted displays, we begin with an overview of the proposed hardware for the Land Warrior helmet-mounted display and a discussion of its intended uses in enhancing soldiers' awareness of their environment. We follow this with the advantages and disadvantages of such displays for the infantry soldier. Of particular concern is that the display may degrade or even block out information about the local environment that is normally available through the unaided eye; it may, because of its weight, reduce mobility; and its use may result in spatial disorientation and dizziness. Next we describe the research base on a series of visual factors to be considered in designing and assessing display devices. These factors include: field of view and resolution, binocular versus monocular viewing, visual perception of the world and pictures, and depth cues. The discussion of depth cues presents a tentative framework for a program of testing, evaluating, and improving military visual displays. We then discuss the value of training in overcoming visual and perceptual distortions. The final section presents our conclusions and design guidelines.

INTRODUCTION

Hardware Configuration

The display in the Land Warrior System is initially to be an opaque screen, with a 40 degree field of view displayed on a monochrome 640 × 480 active matrix electro luminescence (AMEL) display, positioned about 1 inch from the wearer's eye. The display is monocular, leaving one eye available to view the ambient environment. The optics and display are to be suspended from the helmet, with the image intensifier located either on the helmet or in line with the optics.[1]

Several human factors issues need to be considered in evaluating this design. First, there are ergonomic issues related to placing additional weight on the helmet and ensuring that the display is stable with respect to the head; we discuss these issues in detail in Appendix A. In addition, the monocular display, limited resolution coupled with field of view, and off-axis location of sensors have important implications for perceptual and perceptual-motor performance.

Functions of the Helmet-Mounted Display

In the Land Warrior System, helmet-mounted displays are to serve several functions, the most important of which is to display the output of devices designed to enhance soldiers' perception of their environment. These include the night vision system and the thermal weapon sight. The night vision system amplifies ambient illumination and allows soldiers to see night environments that would be essentially invisible to the unaided eye. The thermal weapon sight uses the heat differences between objects and their backgrounds to produce a thermal image of the environment. This image can be useful at night as well as during the day, when smoke and other obscurants can make targets difficult to see with the unaided eye. In addition, the device can display messages regarding danger and troop movements, as well as information useful for navigation, such as maps and location as determined by the global positioning system (GPS).

It should be noted that all of the information listed above can be displayed on devices other than a helmet-mounted display. For example, night vision goggles,

---

[1]The influence of bandwidth constraints on image quality and refresh rate must be accounted for in the proposed wireless transmission system.

thermal sights, and GPS are currently used to good effect by the Army. The potential advantage of the helmet-mounted display is to integrate this information on one display, facilitating rapid switching between various sources of information as circumstances demand. For example, superimposing symbology on the night vision display would allow users to switch back and forth between these two information sources without making head movements, large eye movements, or changes in accommodation. Similar advantages apply to a case in which the soldier must rapidly switch from using the night vision system for movement across a terrain to acquiring a target with the thermal weapon sight.

The use of helmet-mounted and head-up displays in aircraft provides some insights into the potential advantages and disadvantages of this display technology. The Army has pioneered in the application of helmet-mounted displays in aviation, with various night vision devices and sensors feeding into the Integrated Helmet and Display Sighting System (IHADSS). It has developed an extensive body of research data on visual performance with sensor displays (Foyle and Kaiser, 1991; Bennett et al., 1988; O'Donnell et al., 1988), human factors and safety problems (Brickner, 1989; Hart and Brickner, 1987; Rush et al., 1990), and field experience with visual illusions (Crowley, 1991). These analyses demonstrate both the great potential and the risks of using helmet-mounted displays. For example, Wickens and Long (1994) have shown that head-up displays do provide an advantage to pilots in terms of staying on course and instrument landings. However, they have also shown that pilots using a head-up display are more likely to miss occasional, low-probability events, such as an aircraft moving onto the runway during an approach for landing. This may be due partly to the cluttering effect of the symbology's being superimposed on the image of the outside world, as well as attentional conflict between the near and far information domains (Fischer et al., 1980; Hoffman and Mueller, 1994; McAnn et al., 1992; Neisser and Becklin, 1975; Wickens and Long, 1994; Wickens et al., 1993).

The use of helmet-mounted displays by the infantry soldier, however, poses its own particular set of constraints that may be different from those encountered in the cockpit. Because the soldier is mobile, the issue of providing a stable base for the display becomes even more important than it is in the cockpit, making helmet fit and weight critical issues (see Appendix B). In addition, part of the advantage of head-up displays in the aircraft is due to symbology that can be made conformal with various aspects of the scene (Weintraub and Ensing, 1992). A symbolic runway with associated symbology can be superimposed on an actual runway scene, which helps to integrate the two sources of information and reduce attentional interference (Wickens and Andre, 1990). It is difficult to see how this sort of conformal mapping between symbology and scene features could be achieved in the infantry environment. It is therefore important to analyze the use of helmet-mounted displays within the context of the physical and task environments in which the infantry soldier operates.

For example, the Land Warrior System, with its associated soldier radio, may allow a squad to operate in a more dispersed fashion that reduces its vulnerability. However, night use of a monocular display may at times reduce the ability to detect a camouflaged ambush site because of a loss of certain depth cues, such as stereopsis (especially if small head and trunk movements are not made to compensate through parallax for the loss of stereopsis). This problem of target detection is in a sense amplified by the greater speed of movement afforded by the Land Warrior System. Thus, overreliance on the visual system and speed of advancement through the terrain could reduce squad attentiveness to other cues. If an ambush occurs, the squad's speed of execution of the counterambush drill may be reduced because of both the time needed to orient to the enemy and the narrow field of view.

As to the presentation of symbolic data in the Land Warrior System, for some tasks it may be better to place the helmet-mounted display screen off the visual axis or use a hand-held device, which would require the wearer to shift gaze in order to access the information on the screen but might more than balance this cost by reducing clutter. It is important to determine when it is advantageous to present information superimposed on the scene image and when it may be better to provide other displays.

Advantages and Disadvantages of Display Devices

The introduction of devices that provide remote or local information in the form of enhanced sensory or symbolic displays may in the proper circumstances contribute greatly to the safety and effectiveness of the infantry soldier. However, the specific means proposed in each case may interfere with the acquisition or use of sensory information, depending on the circumstances, and all such devices are associated with certain general problems. We introduce such problems briefly here and discuss them in more detail in the next section.

First, as we discuss in this section, helmet-mounted displays may degrade or even nullify information about the nearby environment that is normally available through the unaided senses; see the report on the Soldier Integrated Protective Ensemble (SIPE) (U.S. Department of the Army, 1993). They may distract attention in ways that may have a critical effect on some tasks, interfering with the user's situation awareness. Even design factors that may be unimportant under less demanding conditions may seriously contribute to a soldier's workload under combat conditions.

For example, the SIPE squad and team leaders reported to our panel a situation in which they were unable to see an ambush target even though the target presented itself on multiple occasions. The squad positioned itself further from the kill zone (concentrated area of fire) because they felt secure in their ability to observe the site. One possible explanation of why the squad was unable to detect the target is that their attention was distracted: they reported diligently observing the kill zone, which meant that they were focused on an area. If the target passed outside the area of narrowed attention, they may never have noticed it, even though it was in their field of view.

Second, through excessive fragility, bulk, and weight, the equipment by which remote or sensor information is displayed may seriously reduce the mobility of the dismounted infantry soldier; it may also add fatigue and heat stress (U.S. Department of the Army, 1993). In aviation helmet-mounted display equipment, the heavy and off-center optics increase fatigue and headaches, and the close-fitting helmet liners used to hold the optics in position may also increase heat stress. In the infantry, these physical problems are an even greater cause for concern: greater fatigue can be expected because active infantry soldiers do not receive support from a seat, and the equipment may interfere with their ability to move and to take cover rapidly.

The physical effects of the equipment may have perceptual and cognitive consequences as well. Spatial disorientation can be expected; with no external support, such as a seat, a soldier is not provided with tactile information about bodily orientation to help counteract any disequilibrium due to the helmet-mounted display. Because the weight, weight distribution, and configuration of some displays interfere with the free head movements that a soldier would otherwise rely on to obtain the visual information that is intimately tied to normal action and locomotion in the environment, the equipment-based deficits offered to the infantry would seem to be considerably more serious than those offered in aviation.

These two sets of issues, the sensory/perceptual and the ergonomic, not only are problems in themselves, but also may interact in counterproductive ways. They must be kept in mind by both the equipment designers and the users. Cost-benefit analyses after appropriate testing—tests whose results apply to the situations in which the equipment is to be used—should precede commitment to any mode of proposed enhancements and to the means by which they are achieved. Table 4-1 summarizes the major benefits and costs of key factors of helmet-mounted displays as well as the key research and testing issues.

Before examining visual factors in detail, it is useful to compare the potential side effects of the proposed helmet-mounted display with those found in others currently being developed. Much of the recent work on the effects of helmet-mounted displays has focused on their use in creating virtual environments (VE). In VE applications, the user is emersed in a synthetic environment that differs from the real-world environment. Experiences in VE involve remote synthetic images of scenes, auditory displays, and apparent head and body motion. The current state of the art in VE technology permits display of relatively sparse image geometry (supplemented by "wallpaper texture"), updated at low rates (usually less than 30 Hz), and displayed more often than not in a biocular format. Current head and body tracking systems, which are required to synchronize the displayed scene with user movements, have hysteresis problems, are slow, and are inaccurate at the limits of the operating envelope. The result is often a low-

TABLE 4-1 Display System Features, Human Performance Considerations, and Research Issues

Control or Display Device	Sensory and Ergonomic Considerations	
	Benefits	Costs
Head/helmet-mounted display (general issues)	• Always available • Does not have to be held in the hand or manipulated • Can easily be aligned on target or terrain feature • Wide field of view • Can be used to guide movement • Added information improves situation awareness of medium to long-range environment	• Added weight on head • Off center CG • More complex and fragile than hand-held display • Precision/alignment requirements more severe • Wide field of view results in inadequate resolution • Display information content may overload or distract user, reducing situation awareness

Visual Design Research	Test Approach	Visual Test Conditions	Test Criteria
• Acquisition/application of information on a visual display • terrain • targets • map data • overlaid symbology/ text • overlaid cursors/ reticules • placement (location on display) of information • information density (and clutter) • time for sequencing text, other data • display switching (IR/I^2)	• Laboratory/bench technical test	• Controlled light conditions, controlled display conditions, synthetic images • Assess optical and display parameters (e.g., FOV luminance) • Assess off-axis viewing, distortion, off center display, etc. • Display prototype data formats, realistic targets at varied ranges and aspects to determine peak performance in optimum conditions • Use head tracker to assess search head movements	• Percent correct and time to detect • Identify targets and terrain features • Place reticule on target • Percent correct and time to acquire and apply displayed information
	• Controlled user field experiments	• Measured day/dusk/ night lighting, conditions controlled display conditions, synthetic and real images, real targets at a controlled distance, camouflage, image stability, information legibility while moving, distracting and/or masking effects of HMD on assessing real targets, varied user population to assess peak performance in known conditions	• Percent correct and time to detect • Identify targets and terrain feature • Place reticule on target • Percent correct and time to acquire and apply displayed information • Effects of mobility upon display useability (especially off-axis viewing, interference

TABLE 4-1 Continued

Control or Display Device	Sensory and Ergonomic Considerations	
	Benefits	Costs
Monocular helmet-mounted display	• Minimum weight • Simplest HMD; less alignment required • Eye with no display remains dark adapted • Eye with no display continues to sample real world	• Severe visual rivalry problems. such as target suppression (involuntary) and "cognitive switching" • CG is off sideways as well as forward • Smallest FOV; least information capability; more and larger head movements required • No depth information • Difficulty to navigate on uneven terrain

Visual Design Research	Test Approach	Visual Test Conditions	Test Criteria
		• Assess optical and display parameters (e.g., FOV, luminance) • Assess off-axis viewing, distortion, off center display, etc. • Display prototype data formats, realistic targets at varied ranges and aspects to determine nominal performance in known conditions • Use head tracker to assess search head movements	with real world situation awareness) • Effects of ambient conditions on HMD information delivery, interaction with local environment
	• Operational field testing	• Assess stress, fatigue, varied information content in operational tasks in a field exercise with/against soldiers with conventional equipment	• Effective use of information, success and time to conduct operational tasks dependent upon HMD data, interference of HMD on local SA
• General HMD issues, plus: • Effects of visual rivalry, loss of stereo • Effects of smaller field of vision (FOV) with respect to visual search, reduced information content, more emphasis on format of data	• Laboratory/bench technical test	• As general HMD issues	• As general HMD issues

TABLE 4-1 Continued

Control or Display Device	Sensory and Ergonomic Considerations	
	Benefits	Costs
Biocular helmet-mounted display	• Wider FOV, more information, easier to navigate • No interocular rivalry • Less complex to adjust than binocular	• Heavier than monocular • Poor resolution • Incorrect depth information • Isolates user from environment

Visual Design Research	Test Approach	Visual Test Conditions	Test Criteria
	• Controlled user field experiments	• Assess possible visual fatigue, disorientation, postural stability/loss of coordination	• As general HMD issues plus: • Effects on postural stability, navigational/ vestibular orientation
	• Operational field testing	• Assess stress, fatigue, varied information content in operational tasks in a field exercise with/against soldiers with conventional equipment	• Effective use of information, success and time to conduct operational tasks dependent upon HMD data, interference of HMD on local SA • Effects on orientation, attention fatigue and possible perceptual adaptation with longer-term usage
• General HMD issues plus: • Effects of anomalous stereo/parallax upon target assessment, mobility	• Laboratory/bench technical test	• As general HMD issues	• As general HMD issues
	• Controlled user field experiments	• As general HMD issues • Use head tracker to movements	• As general HMD issues
	• Operational field testing	• Assess stress varied information content in operational tasks in a field exercise with/against soldiers with conventional equipment	• Effective use of information, success and time to conduct operational tasks dependent upon HMD data, interference of HMD on local SA

TABLE 4-1 Continued

Control or Display Device	Sensory and Ergonomic Considerations	
	Benefits	Costs
Binocular helmet-mounted display	• Can provide stereo viewing • Better depth information for mobility • Better target recognition	• Heaviest optics • Alignment and adjustments more complex and critical

Visual Design Research	Test Approach	Visual Test Conditions	Test Criteria
• General HMD issues, plus: • Assess advantages of correct stereo in mobility, target assessment as well as vulnerability of stereo system to misalignment problems	• Laboratory/bench technical test	• As general HMD, plus: • Assess effects of optical misalignment upon performance	• As general HMD
	• Controlled user field experiments • Effects of stereo vision on object detection and and recognition, mobility • Assessment of precision and registration requirements	• As general HMD, issues, plus • Assess effects of optical misalignment upon performance, orientation and postural stability and coordination when moving	• As general HMD issues • Effects of postural ability, navigational/ vestibular orientation
	• Operational field testing	• Assess stress, fatigue, varied information content in operational tasks in a field exercise with/against soldiers with conventional equipment	• Effective use if information, success and time to conduct operational tasks dependent upon HMD data, interference of HMD on local SA • Effects on orientation, attention fatigue and possible perceptual adaptation with longer term usage

TABLE 4-1 Continued

Control or Display Device	Sensory and Ergonomic Considerations	
	Benefits	Costs
Helmet-mounted display with see-through (transparent optics)	• Display content can be integrated with the real world scene • Referenced navigational and targeting data can provide "where to look" guidance • User retains visual contact with the real world	• Display collimation interferes with eye's accommodative response to the real world • Display luminance interferes with eye's luminance adaptation to the real world • Display content may obscure objects in the real world (clutter) • Unstable registration of display image on the real world may induce disorientation
Helmet-mounted display without see-through (world occluded)	• Less complex (lighter) optics • Minor misregistration with real world less noticeable	• User is isolated from real world • Major misregistration with real world is less detectable and can result in serious positioning errors • Significant re-adaptation time to real world when display is removed

Visual Design Research	Test Approach	Visual Test Conditions	Test Criteria
• As general HMD issues, plus: • Effects of display collimation luminance and content upon ability to deal with the local environment, effects of changing display content or moving display content upon postural stability coordination, orientation	• Laboratory/bench technical test	• As general HMD issues, plus: • Assess registration requirements	• As general HMD
	• Controlled user field experiments	• As monocular HMDs	• As monocular HMDs
	• Operational field testing	• Assess stress, fatigue, varied information content in operational tasks in a field exercise with/against soldiers with conventional equipment	• Effective use of information, success and time to conduct operaitonal tasks dependent upon data, interference of HMD on local SA
• As general HMD issues	• Laboratory/bench technical test	• As general HMD issues	• As general HMD issues
	• Controlled user field experiments	• As general HMD	• As general HMD

TABLE 4-1 Continued

Control or Display Device	Sensory and Ergonomic Considerations	
	Benefits	Costs
Helmet-mounted display with integrated symbology and sensor image	• Much more information can be coded symbolically • Critical features (e.g., targets, navigation way points, supply drops) can be localized and enhanced • Remote sensor and intelligence information can be integrated	• Users must be trained to use symbology • Luminance, depth, and apparent size of symbology must be integrated with the sensor image and world • A tendency to load the user with more information than needed avoided • Unstable symbology can induce motion illusions, disorientation, loss of balance

Visual Design Research	Test Approach	Visual Test Conditions	Test Criteria
	• Operational field testing	• Assess stress, fatigue, varied information content in operational tasks in a field exercise with/against soldiers with conventional equipment.	• Effective use of information, success and time to conduct operational tasks dependent upon HMD data, interference of HMD on local SA
• As general HMD, plus: • Effects of display collimation luminance and content upon ability to deal with the local environment, effects of changing display content or moving display content upon postural stability, coordination, orientation • Assess training, time associated with data formats, effects of double imaging on imagery/symbology in overlap regions and/or see through	• Laboratory/bench technical test	• As general HMD, plus • Assess registration requirements	• As general HMD, plus: • Value added, optimum location, interference effect of *each* symbolically coded datum must be assessed in isolation and in conjunction with other display content • Effects of misadjusted symbology luminance, depth, location in perception of the real world • Optimization of information content for specific tasks • Training requirements • Effects of unstable symbology on orientation, mobility

TABLE 4-1 Continued

Control or Display Device	Sensory and Ergonomic Considerations	
	Benefits	Costs

Visual Design Research	Test Approach	Visual Test Conditions	Test Criteria
	• Controlled user field experiments	• As general HMD issues	• As general HMD issues, plus: • Effects on postural stability, navigational/ vestibular orientation • Value added, optimum location, interference effect of *each* symbolically coded datum must be assessed in isolation, and in conjunction with other display content • Effects of misadjusted symbology luminance, depth, location in perception of the real world • Optimization of information content for specific tasks • Training requirements • Effects of unstable symbology
	• Operational field testing	• Assess stress, fatigue, varied information content in operational tasks in a field exercise with/against soldiers with conventional equipment	• Effective use of information, success and time to conduct operational tasks dependent upon HMD data, interference of HMD on local SA

TABLE 4-1 Continued

Control or Display Device	Sensory and Ergonomic Considerations	
	Benefits	Costs
Helmet-mounted display with remote sensor image (e.g., offset sensor, laser sight on weapon)	• Information not locally available may be integrated • Weapons may be aimed without exposure	• Movement with sensors not collocated with the eye can induce motion and position illusions resulting in errors, disonentation, motion sickness • Differences in scale, optical axis, resolution of multiple sources can induce error and confusion

Visual Design Research	Test Approach	Visual Test Conditions	Test Criteria
			• Effects on orientation, attention fatigue and possible perceptual adaptation with longer term usage
• As general HMD issues, plus: • Effects of display collimation luminance and content upon ability to deal with the local environment, effects of changing display content or moving display content upon postural stability, coordination, orientation • Requirements for integrating scale, resolution, optical axis of image sources; assess target/terrain characteristics with non-visual (e.g., thermal) contrast effects • Training requirements associated with using thermal imagery	• Laboratory/bench technical test	• As general HMD issues, plus: • Assess registration requirements • Non-visual wave lengths characteristics synthetic modeled in synthetic imagery	• As general HMD issues.
	• Controlled user field experiments	• As see through HMD issues, plus: • Assess weather effects (temperature, precipitation, fog/haze) which may produce sensor performance variations, thermal camouflage	• As see through HMD

TABLE 4-1 Continued

Control or Display Device	Sensory and Ergonomic Considerations	
	Benefits	Costs

quality image with conflicting sensory cues that become unstable and uncorrelated when the user moves his head. These features of the VE image have led to a number of reported cases of disorientation and motion sickness—a problem that will continue to hamper the widespread acceptance of this technology. In 1992, a special issue of *Presence* highlighted work in this area.

The display proposed for the Land Warrior System is a flip-down, monocular display mounted on one side of the soldier's helmet. During daytime operations, an opaque display will be used to provide navigation information, command and control data, and real-world images acquired through the weapon sight; at night, these functions will be integrated with the night vision system. The proposed displays and optical systems pose some risk with respect to problems such as eyestrain, disorientation, and physical discomfort resulting from ergonomic limitations. The risk of inducing disorientation and motion sickness should be significantly less than for VE systems, however, because there are fundamental differences in the technology. These can be summarized as follows:

- Images to be viewed by the soldier are derived from sensors whereas VE images are generated synthetically. The lags and scene content of the VE system are thus not issues for the proposed infantry system.

Visual Design Research	Test Approach	Visual Test Conditions	Test Criteria
	• Operational field testing	• Assess stress, fatigue, varied information content in operational tasks in a field exercise with/against soldiers with conventional equipment	• Effective use of information, success and time to conduct operational tasks dependent upon HMD data, interference of HMD on local SA • Effects on orientation, attention fatigue and possible perceptual adaptation with longer term usage

• When the displayed image is derived from the night vision sensor, it is correlated with head motion. There are no time lags of the sort that would be induced by a head tracker. As a result, the unstable image problems associated with trackers are not a problem for the proposed infantry system.

• When the displayed image comes from the weapon sight, it is remote and uncorrelated with head movement. However, the pointing direction and rate of movement are directly under the soldier's control so that he can maintain a stable image at a cost in speed of response. Furthermore, when this image is in use, the soldier is stable and braced in a static position.

A poorly fitted or badly balanced helmet will increase the risk of disorientation because, in addition to physical discomfort, the display will be unstable and will move around unpredictably. Other problems with the proposed display include the lack of binocular optics, lag characteristics of the AMEL displays, and the use of the weapon sight image in any situation other than a static brace. However, if the Land Warrior System is well fitted and properly aligned, the risks of motion sickness should be minimal.

VISUAL FACTORS IN DESIGNING AND ASSESSING DISPLAY DEVICES

In connection with night vision in helicopters, Weintraub and Ensing (1992) note that, simply considered, some visual information is obviously better than none. However, that point is less clear-cut if the visual information is misleading; if its correct interpretation requires more training, higher mental capability, and better concentration than soldiers have; and if interpretation takes more time than can be spared under the conditions of use. Such potential limitations on information use must be considered in relation to specific tasks.

An example of difficulty in interpretation and perception was the performance of TOW weapon system gunners employing the first thermal night sight. The TOW missile system had been fielded for several years prior to the development of a thermal night sight. When the night sight was added, it clearly enhanced the system's capabilities and potential; however, there were significant problems in training gunners to detect and identify threat targets at ranges of 1,000 meters and beyond. The night sight was capable of detecting the heat difference at extended ranges, but soldiers had great difficulty consistently detecting and then identifying friendly versus enemy targets. Once detected, the probability of a hit was high, but soldiers could not be easily trained to detect targets effectively and to avoid fratricide under varying battlefield conditions. This training problem was not fully resolved until the quality of the thermal image was improved, and soldiers gained confidence in their ability to discriminate images in the sight.

Visual displays provide layout information through the patterning of light and dark (and color, when applicable) on the surfaces that they present to one eye (monocular), to both eyes (biocular), or by different displays to each eye (binocular). For this discussion, a display with high physical fidelity is one that provides much the same effective patterning to the eye as would the layout or environment itself when viewed under good conditions. Task-independent definitions of fidelity are not now available. When effective fidelity is too low for a specific task to be performed, the display may be useless or even harmful. Because increases in physical fidelity will in the short term entail increased expense, fragility, weight, interference with mobility, and other costs, it is important to achieve an understanding of the effective fidelity needs of different tasks.

Display designs are usually discussed in terms of their sensory properties, that is, in terms of the attributes that the displays can offer to the eye and in terms of which their fidelity limits can be assessed. These sensory properties are readily measured. However, they do not by themselves provide information on whether the display allows the viewer adequate perceptual knowledge of the objects and layout being confronted and adequate situation awareness of the environment in which the actions are to be taken.

TABLE 4-2 Comparison of Field of View Differences for the Human Eye and for Several Input Devices

Eye or Device	Horizontal Field of View in Degrees
Human eye (using both eyes, with no head movement)	210°
Sniper rifle scope (M 49)	2°
Thermal weapon sight (light)	15°
Dragon sight	3.4° to 6.8°
TOW sight (AN/TAS 4A)	8°
BFV driver thermal sight	45°
M113 M-19 driver night sight	26.8°
M 113 vision block (M-17E4)	98° in the 15° uplook position or 76° in the 20° uplook position; (vertical field of view of 23° up and 21° down in the 15° uplook position)
AN/PVS 7 night vision device	40°
Land Warrior System requirement	60° (90° desired)

Field of View and Resolution

Augmentation displays are all vastly impoverished in comparison with the light that reaches the unaided eye from a natural environment under good viewing conditions. Of course, image intensification and thermal imagery provide information that is not available to the unaided eye. Nonetheless, it is important to consider how various choices of display resolution and configuration affect the observer's perceptual abilities.

Sensors generally provide a field of view that is much smaller than is generally available during unaided viewing. Small fields of view are undesirable because the observer loses sensitivity to peripheral information and may have trouble integrating the separate views into a coherent whole. Table 4-2 compares the horizontal field of view of the unaided human eye with that of several visual input devices.

A display's resolution (that is, measured by the number of dots [pixels—horizontal] or by the number of stripes [TV lines—vertical] per degree of visual angle that can be discriminated) is virtually always lower than the normal eye's highest resolution.[2] The display's contrast (the ratio of its darkest and brightest

[2]Other metrics may also be used to define resolution. For example, resolution in night vision goggles is measured in line pairs (see Technical Note).

regions) and the number of intermediate levels between those extremes (if there are any) are far less than the eye normally handles. Approximate guidelines for reading, vernier and stereo acuity, contrast thresholds, and other sensory tasks provide a starting point against which to evaluate the equipment (see Boff and Lincoln, 1988). For the reasons considered below, however, those guidelines are not enough. Tests of the features that permit adequate visual performance under actual or simulated field conditions are needed prior to final design decisions.

The significance of these sensory factors depends on the perceptual and cognitive tasks that a viewer is to perform. At the very least, a display must support the visuomotor actions needed to look at it and retrieve information from it. For example, the display must provide enough contrast to focus the eye on the plane of the display (accommodation) and enough coherence (low enough "snow" or noise level) to allow purposeful eye movements aimed at some peripherally distinguishable feature. Although such behaviors are largely elective and technically under the control of the viewer, their coordinated use has been extremely well practiced in normal environments, as have relationships between the two eyes and between the eyes and the movements of head, body, and limbs.

Binocular Versus Monocular Viewing

A person's two eyes are separated by an interocular distance of 65 mm, and their lines of sight are normally converged to approximately the angle that matches their accommodation distance. For nearby distances (see Table 4-1), each eye receives a noticeably disparate view of the layout. A point in the world that then falls on corresponding places in the two eyes is seen as a single point in binocular vision; a point that falls on noncorresponding places in the two eyes is seen as nearer or further, depending on the disparity. (For more detailed discussion, see Howard and Rogers, 1955, especially pages 55-58). Artificial displays to the two eyes may depart from this natural arrangement in different ways. Each kind of display interferes in some way with this normal process. In monocular displays, one eye is augmented and the other is unoccluded. In biocular displays, both eyes receive the same augmented view. In binocular displays, the two eye's views are disparate so as to provide binocular parallax or stereoscopic depth information, obtained from sensors of fixed vergence.

There are several important issues involved in deciding whether the helmet-mounted display should be monocular or binocular (see Table 4-1). Monocular displays have the advantage of economy and lower weight. In addition, the unoccluded eye is allowed to adapt to the dark and is therefore available for detecting targets in the soldier's immediate vicinity. On the down side, monocular viewing necessarily involves a loss of stereoscopic vision and may lead to binocular rivalry.

Binocular Rivalry

When images shown to the two eyes are sufficiently different, for example, a vertical grating presented to the left eye and a horizontal grating shown to the right eye, the brain is unable to combine the two images together into a single view. A state of rivalry ensues, in which the views provided by the two eyes are seen in alternating fashion. With large images, this rivalry will be piecemeal, with localized patches of one eye's view juxtaposed with patches from the other eye. Several variables influence which eye will dominate this rivalry and how long an image will be suppressed. Generally, the eye with the stronger image will dominate, with stronger images being associated with greater contour, brightness, contrast, and motion (Howard and Rogers, 1995). A closed eye represents an extreme case of low strength, but even a closed eye can sometimes suppress the perception of an eye viewing high contrast, moving lines (Howard, 1959).

In the case of a monocular display, we would expect little rivalry at night because the augmented eye will provide a much stronger image than the unoccluded eye. This would tend to negate the ability of the unoccluded eye to provide information useful for detecting targets. In addition, there should be occasional brief periods when the unoccluded eye gains dominance and causes a degradation of the augmented eye. This would tend to increase in severity with increases in the ambient illumination. Rivalry would be more severe during the day, when the unoccluded eye is receiving a stronger image than the augmented eye, a topic that deserves research under field conditions. Of course, in this situation the observer has the option of closing one eye and reducing rivalry.

An additional factor in rivalry occurs when observers have unequal acuity in the two eyes. According to AR 40-501, Standards of Medical Fitness, infantry personnel must have corrected visual acuity of at least:

1. 20/40 in one eye and 20/70 in the other eye,
2. 20/30 in one eye and 20/100 in the other eye, or
3. 20/20 in one eye and 20/400 in the other eye.

AR 611-201, Enlisted Career Management Fields and Military Occupation Specialty, puts a further restriction on vision requirements for the infantry soldier. To be awarded the military occupational specialty 11B, personnel must have corrected visual acuity of at least 20/20 in one eye and 20/100 in the other eye. An acuity of 20/20 in one eye and 20/40 in the other meets the clinical definition of amblyopia, which occurs in approximately 1-2 percent of the population (Anne Marie Rohaly, personal communication, May, 1996). Amblyopes essentially rely on their good eye for perception. A somewhat similar condition occurs when contact lens wearers use the 'monovision' system, in which one lens corrects for near vision and the other for far vision. Clear vision is obtained at both near and far ranges with the eye providing the clearest image achieving dominance. In the case of amblyopes, the good eye will tend to dominate, and it may be important to measure acuity in each eye before deciding which eye should view the display.

It is also possible that long periods of monocular viewing may result in changes in normal binocular function, although the evidence on this is sparse. Brown et al. (1978) found that, after 8 days of monocular occlusion, subjects showed large changes in phoria and severe diplopia and failed all tests of stereopsis; after 4 hours of occlusion, much smaller and more transient effects have been found (Sethi, 1986).

Stereopsis

In addition to producing rivalry, monocular displays remove the depth cue of stereopsis. Stereopsis is a particularly potent depth cue for objects close to the observer, and the resulting depth sensitivity declines linearly out to a range of about 30 meters (Cutting and Vishton, 1995). Stereopsis is only one of many depth cues that are discussed in a later section, as one can readily demonstrate by closing one eye and noticing that depth information hardly changes at all. This latter observation is sometimes used to claim that stereo is really not that critical to seeing depth. However, it is a bit misleading. First, at close range, stereo depth acuity rivals vernier acuity in terms of sensitivity and can resolve depth differences as small as 2-6 arcsec (Howard and Rogers, 1995). Second, the stereo system operates by matching local features in each eye and can perceive depth in the absence of any recognizable monocular shapes (Julesz, 1971). This ability is especially important in breaking camouflage, in which an object invisible to a single eye stands out in depth against its background. Stereo depth is therefore likely to be particularly important in the infantry soldier's environment, which places a premium on the perception of nearby edges (branches, a ridge, etc.) and objects. At night, under low light conditions (no moon) in which the night vision image is low in contrast, many of the monocular depth cues will not be useful. In this case, a wire strung across a path may blend in with the background but stand out in clear relief with stereo viewing so long as the wire's line has a significant vertical component relative to the retina's axis.

Viewing with two eyes has also been found to be superior to monocular viewing in detecting targets in which depth plays no role. For example, binocular viewing of a threshold-level flash leads to better detection than monocular viewing. Part of this advantage is due to probability summation. If each eye has an independent chance of seeing the target, then two eyes should see better than one (Riggs, 1971). Stereo-blind viewers show precisely the advantage for binocular detection predicted by probability summation. Viewers with normal stereo depth perception show advantages of binocular viewing that are greater than predicted by probability summation, suggesting that they have binocular mechanisms that can sum information from the two eyes prior to detection.

Finally, stereo displays potentially provide a larger field of view than monocular displays although the difference is negligible in current usage. Each eye receives a separate augmented view with an area of central overlap. In the monocular display, the unoccluded eye will have an even larger field of view, but, as pointed out above, at least part of this view may be suppressed by the stronger image in the augmented eye.

Evidence on the Importance of Binocular Displays

These considerations suggest that stereo displays should be superior to monocular displays for seeing depth, moving across terrain, avoiding obstacles, perceiving camouflaged objects, and detecting threshold targets. However, it is important to go beyond laboratory tasks and determine whether these are important factors in performance within the military context. Some indication that they are comes from interviews with Apache pilots of the AH-64 helicopter (Rush et al., 1990). Thermal images of the outside world are presented to one eye, leaving the unoccluded eye dark-adapted and available for seeing instruments and maps in the cockpit. Rush et al. report that pilots sometimes have trouble switching their attention from the bright display to the dark-adapted eye. Some pilots resort to flying for very short intervals with one eye closed, an extremely fatiguing endeavor (p. 14). Practice is reported to be effective in controlling rivalry, but tiring missions apparently make rivalry an additional stressor.

More direct evidence on the importance of binocular displays in the kinds of tasks relevant to the infantry soldier comes from recent studies by CuQlock-Knopp et al. (1994). They had soldiers walk through an off-road terrain while wearing monocular, biocular, and binocular night vision goggles. Independent raters judged performance with the binocular system to be superior to either the monocular or biocular systems, which were equivalent. In addition, the binocular system was preferred by the users. Additional field studies of this type are needed to compare these display configurations in a variety of other tasks, such as target detection.

Both monocular and biocular displays deprive viewers of stereoscopic depth information; all three displays use collimated light, which does not allow accommodation to provide differential focus for objects at different distances. These conditions tend to keep the human accommodation and vergence corrective feedback systems in conflict, resulting (with sustained use) in eyestrain, fatigue, and possibly disorientation (Ebenholtz, 1988, 1992; McCauley, 1984). Although soldiers may be able to adapt to such a system, there may be both short- and long-term costs.

This disorientation could be a significant detractor on the battlefield, rather than a multiplier of capabilities. Infantry School personnel informally reported to the panel that some soldiers had difficulty using the monocular night vision devices. These problems included vomiting, temporary blindness in the unstimulated eye, and temporary total blindness. Similar reports circulated about the

Apache equipment, also informally, and may be reflected in the time limits imposed on its continuous use (Brickner, 1989). Visual rivalry is a major contributing factor that can be reduced only by using the system in an intermittent fashion or by designing a fully synthetic environment (virtual reality). Little is known about the distribution of individual susceptibility to visual rivalry or the long-term health hazards that might be associated with prolonged use. Intermittent short-term use may be a solution, but guidelines such as those developed for Army aviation are needed.

Even without eyestrain, fatigue, and disorientation, limited display resources mean limited information transfer, depending on the task. For example, when field of view, resolution, and gray scale are reduced, fewer different patterns and less information can be displayed, whether the display is an electronically transferred image of the optic array that is faced by a sensor mounted on the helmet or a presentation of maps and computer-generated graphics. This means that a viewer can differentiate and recognize fewer shapes (insignia, equipment, landmarks, etc.) than with normal vision. It means that the static pictorial depth cues seen through the display (the depth cues are basically shapes, revealing interposition, perspective, etc.) are less effective, and so are the behaviors that depend on them.

Reduced field of view also means that: (1) there are fewer objects or parts seen from any position of the head when the wearer is surveying some part of the environment and that there is less context to provide meaning for any detail or object that is in fact seen; (2) a wearer receives less of the ambient or peripheral vision that is important for orientation in the environment; (3) there is less scope for surveying the environment by making fast and economical eye movements (see Table 4-1) while holding a given head position; and (4) there is more need for head movements, which are relatively slow and cumbersome (especially when wearing helmet-mounted displays and associated sensors). Tasks that require rapid scanning of a wide array, as when coming up out of a ground roll, should become impossible to perform smoothly and rapidly by normal perceptual-motor skills.

A perfect system, without any of these limitations, is not currently available for the dismounted infantry soldier; an acceptable trade-off must match the available system to the users' demand characteristics, given the tasks that the users must perform. Table 4-1 presents a summary of trade-offs for visual displays within different categories of visual tasks.

Visual Perception

Visual helmet-mounted displays, and visual displays generally, communicate in at least two different ways: (1) they may present two-dimensional patterns that have meaning for the user and can guide behavior with no need for the viewer to perceive a three-dimensional world from that pattern and (2) they may present two-dimensional patterns that have no meaning in themselves but that act as cues to the viewer's perception of depth, objects, and surfaces in a three-dimensional world. In many actions in the world, it is those three-dimensional perceptions that guide behavior. Although these two channels of information are usually both used in helmet-mounted displays, they are treated differently by virtually all displays, so we consider them separately.

The laser range finder can provide information about the distance of some object in the field, substituting for at least some of the functions served by the visual depth cues, and often surpassing them in accuracy and precision. But the range finder cannot replace the depth cues in providing an integrated perceptual grasp of size, shape, and layout. An attempt to use range to determine some object's size, for example, would require deliberate calculations based on the measured distance and the measured size of the object's image in the display, a time-consuming procedure and one that is probably error-prone compared with humans' very rapid normal perceptual grasp of size and distance. Similarly, a buddy system that uses radio communication to provide triangulation from two viewpoints can offer distance information that may draw on trigonometric calculations to supplement the visual depth cues when the latter are inadequate (or have been defeated by the helmet-mounted display), but it cannot substitute for the rapid and intuitive grasp of the three-dimensional environment that unfettered depth cues afford. The form in which people normally obtain and use the information that underlies human perception of any situation cannot in general be altered without substantial cost in time, error, and situation awareness.

Some tasks require little more visually than that the viewer detect a particular two-dimensional pattern (or classes of two-dimensional patterns) on the display's surface or in the collimated array that it presents to the eye. For example, judging whether an infrared marker in the field of view falls on the pattern projected by a rifle's target requires only that the user judge whether or not two patches of light coincide on the display. For this task, the equipment used in the SIPE project received favorable testimony from its users (U.S. Department of the Army, 1993). Success on such tasks can probably be closely predicted by existing data on the resolution and contrast that are needed to detect contours and points. Similarly, the detection and reading of graphic symbols requires only that the display's contrast, resolution, and signal-to-noise ratio fit what is known about legibility (Helander and Rupp, 1984; Grandjean, 1987; Human Factors Society, 1988). There are probably many tasks that can be reduced to similar visual questions about the display's surface. The substantial psychophysical research literature on visual detection and research on computer and video displays can provide good bases for design trade-offs (Helander and Rupp, 1984; Grandjean, 1987; Human Factors Society, 1988).

However, most of the visual tasks that are normally required of infantry soldiers do not depend on information about what contours and points can be distinguished on the surface of a display. They depend on recognizing which three-dimensional objects in the environment need action, on perceiving their three-dimensional spatial relationships to each other and to the viewer, and on perceiving the three-dimensional environment in which the actions are to be performed. Those perceptions depend, in turn, on depth cues, which are patterns of light either projected to the eye by the environment or transmitted through the helmet-mounted display. Even when the display shows only maps or diagrams, which can in principle be provided as two-dimensional patterns and evaluated as such, the literature suggests that the information about layout and location is still best conveyed by the same kind of depth cues that are used in pictures of three-dimensional space (Bemis et al., 1988; Burnett and Barfield, 1991).

Depth Cues

For visually guided behaviors in three-dimensional space (advancing, aiming, reconnoitering or scanning the environment, combat, manipulating equipment, etc.), a viewer can normally draw on various kinds of depth information available to unobstructed vision. Chief among these are pictorial depth cues, motion dependent cues, and cues that involve the adjustment of the ocular system, as we discuss at some length in this section. For general references see Cutting and Vishton (1995), Gillam (1995), and Hochberg and Brooks (1996); the interaction of different sources of information in any specific combination of task and environment must be separately addressed.

Pictorial depth cues do not rely on moving pictures. They are essentially patterns in a two-dimensional array of light that the world projects to the eye, patterns that are themselves two-dimensional but that most frequently arise from differences in depth. That is, the various cues are aspects or features of the pattern of light that are likely to be provided both by a three-dimensional scene (layout of objects and surfaces) and by the projective picture of that scene as made by a camera, by an artist, or by a computer. Thus, because the visual angle subtended by an object decreases with its distance, one has linear perspective (e.g., within the display, sizes perpendicular to the line of sight decrease as distance increases) and the related cues of projective relative size (the relative distance of two objects in the world are inversely related to the sizes of their images in the display) and textural density (as distance in the world increases, the density of textural detail in the image of a homogeneous surface increases). Height in field is a powerful depth cue when viewer and target object rest on the same plane (the farther the target, the higher toward the horizon line). Interposition is an exceptionally strong cue whenever the images of two objects overlap (when two contours form a "t," the uninterrupted one is probably the nearer).

Other important depth cues result from motion parallax, in which characteristic patterns of motions in the two-dimensional array normally arise from the relative motion between viewer and environment (motion perspective, the optical expansion pattern, etc.).

Finally, there are the depth cues that depend on the adjustments of ocular musculature: stereopsis and accommodation. In stereopsis, or binocular disparity, any point in space on which the two eyes are converged (a voluntary act) falls on corresponding points in the two eyes; any object or point at some other distance near that line of sight projects disparate images to the two eyes. Those disparities, taken with the eyes' vergence, are particularly powerful depth cues for relatively nearby distances (Alpern, 1971). Also, the eye muscles increase the lenses' curvature to focus on nearer objects; such muscular action, or accommodation, is a depth cue within very close distances.

Quite different two-dimensional patterns can thus act as cues to the same three-dimensional situation, providing the same perceived depth. It is normally the perceived three-dimensional situation that determines action and decision, when the viewing conditions permit. Even though they are used to provide information that may be missing from the normal field of view (with dark of night being the extreme case), helmet-mounted displays will generally degrade these depth cues; depending on their design, they will do so to a different extent and in different ways.

For example, where helmet-mounted displays transmit light originating in the scene, collimation places all points at infinity as far as accommodation is concerned. Resolution, gray scale, and field of view of the transmitted or reconstructed image are always impaired relative to unaided vision; this degrades those cues that depend on detail (like textural gradients), on gradations of shade (like modeling), or on expanse (like linear perspective). Other cues used for perceiving the spatial layout of objects and surfaces can be generated by computers. In principle, therefore (although it is not contemplated in the programs considered here), helmet-mounted displays might restore or enhance depth information, might use depth cues to provide simulated environments, or, much more modestly, might use depth cues to enhance the separation of different sets of alphanumeric or graphical data.

The effects of degrading or enhancing depth cues, and the trade-offs involved, can probably be estimated. To do so, one would need to assemble a set of matrices around the four variables: (1) which distances are most important for particular tasks, (2) the distance ranges over which each class of depth cues is effective, (3) which depth cues are offered by particular environments, and (4) the effects of different display properties on the different depth cues. A first step toward achieving item (2) has already been taken by Cutting and his colleagues (Cutting and Vishton, 1995); the other items can probably be approximated. Such attempts are necessarily still quite speculative, since at present far less is known in an engineering sense about how the physical properties of a display affect the effectiveness of the depth cues than is known about the display of two-dimensional patterns. They would, however, suggest on a principled basis what research and testing are needed for different tasks and conditions. A discussion of a few examples here illustrates the relationship between these variables and their importance to the design and use of helmet-mounted displays.

First, tasks differ in the ranges of depth information they require. Information about nearby depth (i.e., within 3 meters) is needed in reaching for tools or weapons, for avoiding collision with obstacles (a nearby doorjamb, tree trunk, etc.), and for touching pen to map. Information about whether a moving figure is about to disappear behind a house or wall, or will remain in front of it, involves intermediate distance (say, from 5 meters to 3 kilometers); the same question about a column of moving vehicles and a rise in terrain involves far distances (say, above 5 kilometers). Distinguishing a camouflaged object from its background at any of these distances requires depth information appropriate to the object's scale and distance. If a helmet-mounted display design significantly degrades the depth cues that normally provide the information for depth perception, as most such devices do, performance of any tasks that need such information will take longer and be less accurate.

The Helmet-Mounted Display and Task Performance

In evaluating the proposed helmet-mounted display and sensors, it would be useful to have some idea of what kinds of visual tasks will be affected by limits in display resolution, field of view, gray scale, etc. However, predicting with confidence what will be recognizable in any situation depends on having a usable model, or generalizing from previous tests in that situation, or both. At our present state of knowledge, neither is sufficiently reliable or complete, and so we must generally use both.

Several recent discussions of models intended to apply to electronic instruments are found in Peli (1995). Models that attempt to take into account how information in the target scene is transformed by the electronic media, the visual optics, and selective retinal sensitivities have increased in sophistication, graduating from point-by-point analyses (from pixel to blur circle to retinal spacing) to modulation transfer functions and numbers of cycles per target, and even numbers of cycles per the details needed to identify a particular vehicle (O'Kane in Peli, 1995). These approaches attempt to model target detection and identification in terms of variables such as contrast and spatial frequency. However, it seems likely that accurate prediction will ultimately have to specify the features or shape primitives that underlie shape recognition by humans. These primitives are not currently known, although there are several promising proposals, such as Biederman's geon theory (Biederman, 1995), that have opened promising avenues of inquiry (Ullman, 1996). In addition, it is clear that top-down effects of familiarity, and priming from prior glimpses, permit extremely efficient search and object recognition, but these factors are at present very difficult to model.

Without an explicit or implicit model, laboratory tests are difficult to apply. In any case, laboratory tests alone are of questionable validity, given the many ways in which field tests may differ from the laboratory; indeed, even individual field tests are of questionable validity if the results are to be applied to other situations, to users with other training, to different tasks, etc. (see O'Kane, 1995). Test conditions should come as close as possible to the situations for which the equipment is intended; the tests should come as early as possible in deciding the costs and benefits of different equipment specifications; and any help from operational units in determining what specifications are needed for different tasks would be welcome.

We have approached the task of estimating the impact of the helmet-mounted display on soldier performance by first recognizing that soldiers have to be able to perform visual tasks at a wide range of distances using a varied set of depth cues. The classification recently undertaken by Cutting and Vishton (1995) estimates the importance of various depth cues in perceiving depth over three different ranges or *action zones*. We can use this classification, together with a set of activities a soldier might have to perform using the Land Warrior System equipment, to estimate what kinds of tasks would be affected by limits in display resolution. These estimates are obviously very rough approximations, but they nonetheless may be useful as heuristics.

Depth Cues Used to Guide Action

There are many different channels (or cues) through which information can be obtained about depth and distances in the environment. These differ in substance and mechanism and are unified only by the fact that the diverse channels bring information from the same world, agreeing with each other to the extent that they are all correct. Although such channels are redundant, they do not measure the same things about layout and are not effective over the same ranges. They therefore differ in terms of which task performance they best support. They are differently affected by the devices that reduce resolution, field of view, binocularity, and free head movement; a detailed study of such differences would probably help as a guide both to actual testing and to training. In what follows we (1) introduce a first pass, based on Cutting and Vishton (1995), at classifying the various major cues in terms of three categories or zones of performance in normal perceptual activity and (2) consider for each of these three zones the likely effects of impoverished and offset displays on the wearer's perception of depth, layout, and orientation and what such an examination suggests in the way of necessary further analysis, research, and training.

Visual Guidance within Three Zones of Action

Figure 4-1 should be regarded as a tentative first step in constructing a framework for a program of testing, evaluating, and improving military visual displays. More data and more extensive analyses are needed to solidify and fill

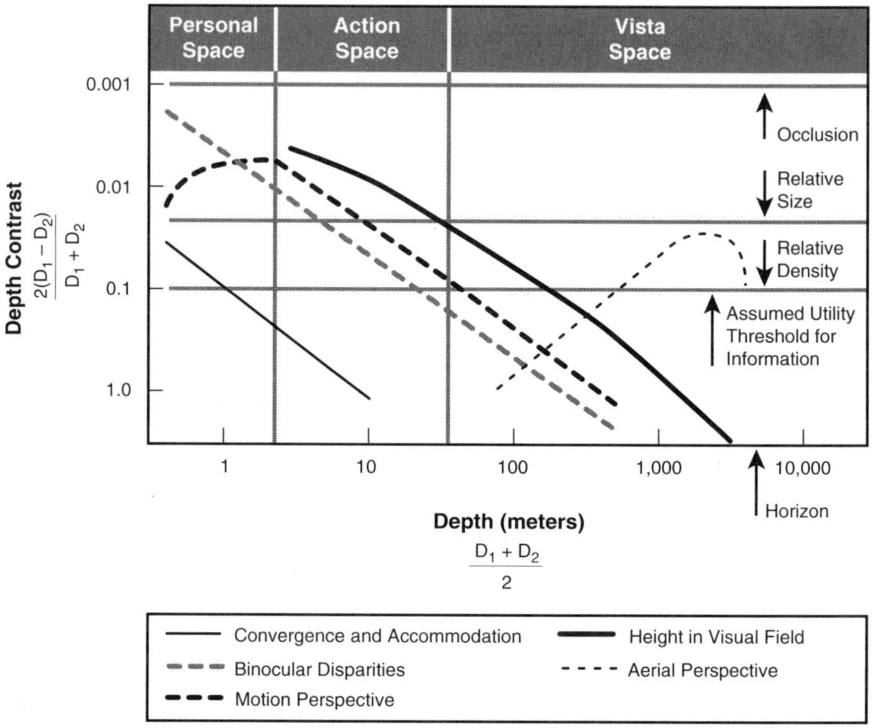

FIGURE 4-1 A framework for testing visual displays in three zones of space. Source: Adapted from Cutting and Vishton (1995). Printed by permission.

out the details of the relationships it proposes. Even more important, it must be expanded greatly if it is to help in considering the effects and aftereffects of the specific displays that are to be used in the various missions of dismounted infantry soldiers. When so expanded, it must be studied in real or simulated field tests to determine how best to deal with display limitations and to determine the specific cautions and training such effects and aftereffects make necessary. There is not now, and is never likely to be, a way of generating the answer to such questions from first principles or from some look-up table, but the body of perceptual knowledge we now have makes it possible to suggest what should be looked for and what will be found. What follows should be considered sample suggestions; they should not be taken as exhaustive.

The plots in Figure 4-1 suggest that there are three egocentric regions or zones of space: Zone 1 is personal space, extending to slightly beyond arm's reach, and delimits the space used by a static observer. Zone 2 is action space, extending to about 30 meters, and encompasses distances in which an observer can throw an object to another person, throw an object at an animal, or easily talk to others. Zone 3 is vista space, extending beyond 30 meters, and includes an area in which changes in the position of objects is slow as a pedestrian moves through the environment.

For each zone, we list the aftereffects to be expected, as well as the effects that impoverished and offset artificial displays are likely to have on picking up spatial cues, apprehending the shapes and layout of the viewer's environment, carrying out typical actions, and perceiving one's orientation in the environment.

For each of 9 depth cues of the more than 15 depth cues that have been widely discussed and studied, Cutting and Vishton use existing data or plausible assumptions to estimate the barely discriminable (threshold) depth separation between two objects at different distances from the designated viewer D1 and D2. These distance are used to derive a measure of depth-contrast sensitivity (DCS) according to the equation: $DCS = 2(D1 - D2)/(D1 + D2)$. Figure 4-1 shows this sensitivity measure as a function of the objects' mean distance from the viewer $(D1 + D2)/2$). The horizontal line at .1 on the sensitivity scale represents the "assumed utility threshold for information." That is, Cutting and Vishton assume that depth differences less than this amount do not contribute to perception of layout. Note that small values of depth-contrast sensitivity reflect good sensitivity. For example, occlusion provides extremely fine discrimination of which object is closer and is effective at all distances from the observer for which the objects' junctions with other objects are visible. By comparison, convergence and accommodation are useful primarily at distances less than 10 meters.

Visual Guidance Within Zone 1 Insofar as personal space is concerned, accommodation and convergence are potentially at their most useful with normal vision, but accommodation is anomalous as a differential depth cue with collimated displays, and convergence is typically fixed by the optical design and therefore anomalous as a cue as well. Users of these devices should be thoroughly warned and convinced that they have lost the normally automatic depth knowledge that is based on this information, and they should be trained to use other cues. In addition, it appears that collimation does not necessarily cause observers to accommodate for infinity. Iavecchia et al. (1988) reported that most observers tend to let their accommodation "lapse inward" when viewing collimated displays. Edgar et al. (1995) recently confirmed this finding and showed that it was especially pronounced when observers had to make complex discriminations of helmet-mounted display imagery. The effect of this incorrect accommodation would be to blur the image as well as to affect the perceived size and distance of objects.

Figure 4-1 shows that occlusion is a highly effective depth cue for all zones, including Zone 1. However, low display resolution can seriously degrade the effectiveness of this depth cue in two ways: (1) The depth information potentially offered where the boundaries of the occluded and occluding objects intersect can normally rest on very fine detail (in normal vision, the threshold for misalignment can be as low as 1 arc sec) and in that case will be lost with the low resolution characteristic of electronic displays, such as the image intensifier (I^2) device and the thermal weapon sight. Depending on configuration, ambiguity of what depth is in fact perceived and even illusions can result (Chapanis and McCleary, 1953; Dinnerstein and Wertheiman, 1957). Users should be alerted to this fact and informed that moving the head laterally while maintaining fixation on the likely region of intersection should help to alleviate this problem; training will probably be needed because the weight of the headgear may discourage lateral head motion.

Overall shape is also normally used to interpolate the region of the intersection cue supporting occlusion (see Chapanis and McCleary, 1953). In a cluttered environment, however, viewers may not discern where one shape ends and another begins. In normal vision, surface quality, color, and texture probably serve heavily in this regard, but low resolution devices lose the texture and sparse gray scales lose the shading. Of the two, it is probably most effective (and certainly cheaper) to increase the gray scale and gamma, so that surfaces are distinguished as much as possible by display luminance. Again, training in the use of lateral head movements will help, and any differential displacement of the light source (as would then occur if the infrared sensor were head-mounted) could also prove useful.

Binocular disparities are also at their most effective within Zone 1, and the viewer is completely deprived of these when monocular or biocular displays are used. It is likely that, if the observer makes small lateral head movements, equivalent information can be regained (albeit more slowly and much less instinctively). We emphasize the need for head-movement training because it seems likely that most tasks to be performed within Zone 1 use focused and relatively fixed attention. Even in normal conditions (aside from something done very close up, like threading a needle), this is probably accomplished with a relatively fixed head and with great reliance on the binocular cues. For that reason, we do not expect that reduced field of view will be much of a problem in this zone. For the same reason, however, the novice will probably also need practice in making the required movements habitual in disambiguating the layout at hand, while ceasing such movements when they would interfere with fixed and tightly focused attention. Although a little informal testing should convince the viewer that elective head movements can be brought into service in this way, we know of no actual research on this matter.

The necessity of adapting visuomotor performance (including the head movements mentioned above) to the effects of offset, or displaced viewpoint, should be greatest in this zone. Wearers should be alerted to the nature of the adaptation, to its incompleteness, and to the fact that they should expect involuntary aftereffects to follow any prolonged use of offset displays.

Visual Guidance Within Zone 2 Occlusion remains useful in Zone 2. Its vulnerability to low resolution could be somewhat lower if the larger objects at issue in action space are smoother shapes (lower perimeter/area ratios and therefore less likely to provide conflicts between the global and local components of occlusion noted above). Their greater separations might make luminance differences more likely within the display, but gray scale will probably continue to be important. Lateral head movements would in most cases have to be impracticably larger in this zone, in order for motion parallax to disambiguate the occlusion with the displays being planned, and therefore training should probably reflect this fact.

Accommodation, convergence, and binocular disparities are anomalous when using the proposed devices in Zone 2. It seems likely that the absence of the binocular information is particularly costly in this space, and training should stress the fact that errors are likely to arise from the monocular viewing (probably mainly ones of assuming that objects that are adjacent in the displayed image array are at the same distance from the viewer). As noted above, one cannot expect that comfortable lateral head movements can serve as useful correctives, but lateral body movements or sideways steps should usually serve to disambiguate the layout. Height in visual field is probably a highly useful potential cue in this zone. At this point, however, we should consider the major problems that will surely be introduced by what we have called the viewpoint offset provided by the equipment and that should be addressed in research and training.

A helmet-mounted sensor changes height relative to the ground plane and horizon and also must distort the viewer's perception of the upright (and slants of all surfaces) (Held et al., 1975; Johansson and Börjesson, 1990; Matin and Li, 1992; Proffitt et al., 1995). These are probably minor effects except when shifting between the display and a direct view; the user's sense of the vertical will have shifted when adapting to the offset, and an opposing shift should occur as an aftereffect when shifting to direct viewing. Practice in anticipating such aftereffects may be profitable. A gun-mounted sensor such as the thermal weapon sight, if used for anything other than centering the target, is very much more likely to distort perceived height in field and badly confuse the viewer's perceived slant of upright and ground plane.

Moreover, since the user must integrate information over changing views as the gun's direction changes without the proprioceptive feedback that would usually provide the context for head and trunk movements, the relationship between those views may be badly distorted or chaotic. That is, because of the restricted field of view, the displacements within the visual field will not necessarily be correctly or unambiguously interpretable. This is known in the computer-vision literature as the aperture problem. It can result in extremely robust illusions as to the direction of motion (often known as the "barber pole" illusions). This in turn may scramble the perception of successively viewed objects' lateral spatial relationships.

Even setting aside the gun sight as a source of layout information, the problem remains in diminished form when any offset sensor is used with reduced field of view. Although occlusion may be a potentially useful depth cue for normal vision in Zone 2, it may be highly vulnerable because of restricted field of view. It seems likely that training to achieve a disciplined use of landmarks and of deliberate panning movements may be able to minimize damage from the aperture effect.

Relative size and familiar size are theoretically both valuable depth cues with normal vision in Zone 2. Because relative size depends on the difference in size of the images provided by two objects at different distances in space, and because there is no reason to believe that it remains effective if the field of view is too small to display both objects simultaneously, relative size may not function reliably with the equipment being considered. This particular issue can probably be tested most cheaply with laboratory simulations. Familiar size presumably provides absolute distance information with a single object and should not therefore be so dependent on a large field of view, but it may in fact also be unreliable and slow even under normal vision.

This raises another problem that is more serious than loss of depth perception: for familiar size to work as a cue, the object must be recognized, and the low resolution and sparse gray scale of the proposed display may interfere with recognition of all but the most distinctive forms. For relative size to work, things that are of very different shape in the world should not end up having similar shape in the display. This is likely to be a problem with the low-resolution images provided by infrared imagery in which shapes may not be recognized and many of the remaining depth cues, such as texture gradients, shading, etc., may not be available. Whether or not this is a problem for familiar size and relative size as depth cues, or for the even more important tasks that hang on object identification, should be tested for specific missions and equipment.

Viewpoint offset probably requires less adaptation in Zone 2 than in Zone 1, and aftereffects are probably correspondingly less disruptive; training here may not be necessary, although that should be determined by research over the appropriate distances. The effects of reduced field of view on integrating an overall picture of the environment should in general be mixed, because larger stretches of the environment are included within the same visual angle, and stable landmarks (large in the world but relatively small within the display) should help the viewer integrate the views obtained from different directions. This requires some degree of shape fidelity between object and display, and we have noted above that, with low resolution and sparse gray scale, this may pose a problem. As with offset, directed research is needed to estimate the extent of the problem.

Visual Guidance Within Zone 3 As Figure 4-1 indicates, in Zone 3, or vista space, binocular disparities contribute little. Infantry soldiers do not normally move far and fast enough to provide useful motion parallax, leaving occlusion, texture gradients (relative density), height in field, aerial perspective, and relative size as more or less effective depth cues for normal daytime vision in much of this space. Visual displays that reduce resolution, gray scale, and field of view should have mixed effects on these cues: relative size and height in field may be damaged less in Zone 3 than in Zone 2, because two relatively distant objects that are not too far apart laterally within the environment are more likely to be simultaneously included within a limited field of view than are two nearer objects. Textural gradients are almost certainly lost with low resolution, and aerial perspective should be rendered useless by eight-bit gray scales.

It is probably not necessary to assess further the effects that such displays have on daytime vision because we assume that (regardless of what designers now intend) most visual tasks during the day will be conducted with unaided vision. At night, moreover, using unaided vision, we should probably expect that Zone 3 will effectively fail to provide information about layout and depth, so that the viewer will have to depend on instrument-provided range information about those distances. Under these conditions and considering the large distances involved, concern about viewpoint offset and corresponding aftereffects does not seem justified. However, the integration of successive views of the sparsely populated nighttime vistas available with small fields of view probably requires both training and the addition to the display of some suitable directional framework, a framework that is salient but that does not obscure the already-restricted visual field.

Effects of Degradation of Depth Cues

Using the above classification, together with a set of activities a soldier might have to perform using the Land Warrior equipment, we now estimate what kinds of tasks would be affected by limits in display resolution.

Image intensifiers present essentially photographic images to the eye, although heavily modified by range and atmospheric conditions and limited by resolution, field of view, and sensor offset. In contrast, thermal imagers depend on characteristic signatures and hot spots; although there are models that seem to predict performance to some extent (O'Kane, 1995), we do not discuss such imagers here.

First, taking several examples from each of the three action zones, we make exceedingly rough estimates of the resolution needed to execute typical tasks in those zones, using equipment that provides the graphic equivalent of image-intensifying devices (Table 4-3). (Although the Land Warrior equipment has been forecast as having higher resolution and larger field of view, the panel was unable to obtain specifications, in any case, the same sorts of analyses will be needed with whatever parameters are finally achieved.) We propose that, even if it is necessary to limit display size and processing load to the present 300 K pixels, the user should have available one or two alternatives to the uniformly distributed 640 × 480 format. Specifically, the user should be able to choose a higher resolution in the lower part of the field, at the expense of the upper part.

TABLE 4-3 Rankings of Information Sources by the Areas under Their Curves in Figure 4-1 within the Three Kinds of Space

Source of Information	Action Personal Space	Space All Sources	Space Pictorial Sources	Vista Space
1. Occlusion and interposition	1	1	1	1
2. Relative size	4	3.5	3	2
3. Relative density	7	6	4	4.5
4. Height in visual field and height in the picture plane	—[a]	2	2	3
5. Aerial perspective and atmospheric perspective	8	7	5	4.5
6. Motion perspective and motion parallax	3	3.5	—	6
7. Convergence	5.5	8.5	—	8.5
8. Accommodation	5.5	8.5	—	8.5
9. Binocular, disparity, stereopsis, and diplopia	2	5	—	7

[a]Dashes indicate data not applicable to source.

Assuming that the processing limits cannot be exceeded, adding additional sensitive units to the sensor and switching them in and out as desired (trading off density in one region against another) would be one way to achieve this, and probably relatively inexpensive. In general, there are different kinds of evidence that the lower part of the visual field is more important for detailed functions, and that the visual system is equipped with more specialized contour-sensitive mechanisms, than the upper field (Previc, 1996; Rubin et al., 1996). More specifically, as we discuss below, depth perception and manipulation, especially in the absence of binocular vision, is not well served by the existing resolutions.

In addition, the restricted field of view (30-40 degrees) could be quite dangerous, because of its negative effects both on situation awareness and on stitching together successive narrow glances at an active and cluttered environment (see examples 1 and 2 below). The 640 × 480 array resolution, although sparse for the central 4-8 degrees of central vision, is much higher than is needed for peripheral vision (the ambient system), and some redistribution should make it possible to increase the field of view to something between 50 and 60 degrees by lowering peripheral resolution. (Luminance modulation could be used to obtain higher effective subpixel resolution; such enhancement might help for some environmental tasks and detract in others, so that real and simulated field tests are important.) The additional margin will not help in obtaining information through eye movements, but it should prove useful when the soldier relies on head movements, as when walking or surveying the surroundings. This display format should probably be elective.

Zone 1 Example 1: Disarming mines, cutting wires, adjusting sights, applying first aid, setting fuses, clearing weapon malfunction, etc. Two issues are detection and depth perception (and aligning parts in depth as well as in the frontoparallel plane).

- Close detail. Assume the following situation: a distance of about 0.5 meter, wires or other parts as small as 1/8 inch, and horizontal display lines (or pixel rows) of about 0.063 degree = 3.75 min (30 degree 480 lines). The 1/8 inch subtends 23 minutes, or is about 8 pixels high and is well above both the 1 min minimum separable for spatial resolution at adequate contrast and the pixel size (3.75 min/pixel) limit of the Land Warrior System. Given the parameters, such features should be visible almost out to 1 meter, at which point the pixel size should exceed the image size, and visual confounding and loss of detail should become a factor.
- Depth localization and 3D form. With only monocular viewing, the depth perception needed to align parts in the third dimension would most naturally come from small head movements. With lateral head movements of about 1 inch, depth differences of about 1/4 inch would be needed at 18 inch (0.5 meter) distance, and about 1.25 inches at 1 yard/meter. Most fine manipulation tasks are conducted within this range or a bit less, and this resolution would likely limit task precision. With twice the resolution in the lower half of the display field, all of these tasks would probably be feasible at 0.5 meter and some even at 1 meter.

Example 2: hand-to-hand combat, breach obstacles, detect branches and handholds, operate controls. At a range of about 1 meter, the field of view should be less than 2 ft. Limbs, weapons, and branches are safely above spatial resolution, but shoulders, limbs, and most of the target body fall beyond the field of view. Something like a 55 degree field of view would include the opponent's head, shoulders, and arms and at a half-normal resolution should then be enough for the ambient system.

Zones 2 and 3 A critical activity occurring in Zones 2 and 3 involves object detection, recognition, and identification (see Technical Note). A soldier needs to detect the presence of another person in the distance, recognize that person as friend or foe, and in some cases identify the individual. The bases for these object recognition tasks are unknown, but we attempt to use some simplifying ideas, similar to the basis of Biederman's geon theory, to provide estimates of what kind of performance might be expected using limited resolution displays. The main assumption of the geon theory is that object recognition is accomplished by recognizing combinations of a small set of component forms. We have no reason to believe that there is a single system underlying object recognition or a single set of criteria. Indeed, it seems reasonable to assume that different classes of objects are recognized and/or identified by different criteria in different zones.

Viewed from 1 meter or less, with a field of view of 30 degrees, a tank's overall silhouette would seem unobtainable. Conversely, viewed from 75 meters, the largest of some soldier's component features (e.g., a 1 inch nose, seen in profile) subtends little more than 1.3 min, whereas the minimum separable angle for reading letters is taken as 1.0 min and, more to the present point, the minimum pixel size is close to 4 min. To pick up the contribution of the nose is barely possible with good unaided vision alone at 75 meters and is not possible with a device of the Land Warrior System's resolution. This is not to say that discerning the target's nose is necessary or sufficient to recognize the soldier (some other feature or clusters of features may be needed). However, if the presence of the largest feature (whatever it may be) cannot be detected, then smaller ones become irrelevant.

Having considered a few examples of resolution effects on specific tasks, we turn to a more general survey of the effects of helmet-mounted displays on depth information.

Effects of Helmet-Mounted Displays on Depth Information

Helmet-mounted displays, like night vision devices, capture optical information about the environment and present it visually to the wearer. If binocular sensors are used, stereopsis may be enhanced (if somewhat distorted) by increasing the separation between them, extending depth information from Zone 1 into Zone 2 (i.e., intermediate distances). If only a single sensor is used, stereopsis is necessarily lost with monocular or biocular head-mounted displays, and accommodation loses all differential depth information in all displays because the light is collimated. Head-motion parallax is seriously distorted whenever the sensors are at a different optical location from the eyes (as in virtually all head-mounted displays) and is lost when a viewer must remain stationary. Without stereopsis or parallax, a viewer is left only with interposition for nearby depth information, and that information is severely limited; objects' images must be overlapping to provide it, and at most it tells only which surface is the nearer.

Because a viewer who must remain stationary but who is concerned more with intermediate and far distances than with near distances must in any case depend chiefly on the pictorial depth cues (see Figure 4-1), the loss of stereopsis and the distortion of head-motion parallax that is imposed by most helmet-mounted displays may not represent significant *additional* costs. That can only be true, however, if the equipment provides the depth cues in a condition that is adequate to the perceptual needs of the task. Such devices are not equal to what the unaided eye receives under good viewing conditions, but more graded assessments are needed for making design decisions, particularly under degraded viewing conditions.

There are data that offer some guidance as to the equipment needed for stereopsis (in the case of binocular displays) and for parallax, obtained as functions of luminance, contrast, and resolution, using very simple displays involving wires, dots, and gratings (Schor, 1987; Schor et al., 1984, Foley, 1987). The pictorial depth cues, however, are another matter. At the most basic level—as two-dimensional patterns—there are various data showing that the exposure time and contrast needed to detect targets vary with their size, background luminance, etc. These could be applied to features of the depth cues that seem amenable to such analysis.

For example, textural gradient and local occlusion may be presented to the sensor by the environment, but they are probably very readily lost at low resolutions; shading cues and the intersections that provide interposition information are degraded or lost with a sparse gray scale; height in field and convergent linear perspective, both of which depend on some extended region of the display, must at some point be degraded as field of view decreases, as aperture-viewing studies confirm (Hart and Brickner, 1987). The effects on these attributes of any display must therefore be assessed before deciding to use the equipment in any task that is likely to call heavily on those cues, and effort should be made to relate, extend, test, and apply the results of such task-oriented analyses.

The same issue arises in the context of computer-generated graphics. One function of the display equipment (including hand-held displays, which do not directly interfere with normal visual perception of the world) is to present maps, diagrams, and charts. Experimental evidence suggests that maps and diagrams that incorporate depth cues may be more effective than traditional displays (see, e.g., Bemis et al., 1988; Burnett and Barfield, 1991), depending on the task and on the parameters and combinations of cues (Ellis et al., 1991; McGreevy and Ellis, 1986). The pictorial depth cues (and stereopsis and motion-based depth information as well) can, in principle, be successfully constructed on computer graphics displays, but they can be used by a viewer only to the degree that the sensory qualities of the display permit. The enhancement and simplification techniques that are available to computer-generated images can provide more robust information, but limitations in field devices such as resolution, contrast, and field of view must still be evaluated as to their effects on depth cues from these artificial sources.

Most familiar classes of objects can be recognized with extreme rapidity even in the absence of any depth information, cued only by their shapes in the display (Biederman, 1985; Peterson et al., 1991). Performance tasks that do not require any specific depth perception—but that can be carried out by recognizing some object(s) or layout (e.g., the presence or direction of a person or group, of a particular kind of equipment, a particular house, etc.)—are therefore probably not badly degraded by the absence of depth information. Moreover, objects' familiar sizes can actually act as depth cues, although such distance information takes longer to extract (Predebon, 1992). Like the depth cues, however, the perception of the objects as shapes necessarily depends on the quality of the display, the field of view, and how the situation prepares or primes the viewer.

In a display that is too coarse to resolve the features that characterize a given object, or with gray scale and contrast inadequate to model its forms, shapes may not be recognizable. Conservative estimates of what tasks can be performed are certainly possible to make (see discussion on action zones). But object recognition cannot be predicted solely from any table of data because objects (and depth cues as well) are normally highly redundant. That is, a part or a feature (or even just an attribute of some object, like its color) may serve instead of the whole object, given an appropriate past or present context (Hochberg, 1980). With training and familiarity, therefore, performance may surpass what would be expected from such tables. Conversely, a given point of view may obscure the relevant features even when display quality is otherwise adequate. As a consequence, trade-off assumptions about any specific equipment need to be tested in the real or simulated missions for which it is intended.

Effects of Helmet-Mounted Displays on Field of View

With a small field of view, some or all of the redundancy in an object or scene is very likely to be lost, because only some portion of either may be included within the display. Moreover, peripheral vision is greatly decreased by the field of view for most displays proposed. This will interfere with object recognition, because the information in peripheral vision normally informs a viewer where to look next in order to obtain some needed feature. In the division of perceptual labor between what have been called the *ambient* and *focal* visual systems (Hughes et al., 1996; Leibowitz et al., 1982; Schneider, 1969; Trevarthen, 1968), which are roughly equivalent to peripheral and foveal vision, it is the former that contributes most heavily to orienting (e.g., attentional capture), visual guidance of the limbs, and posture (orientation or vection) (for recent reviews, see Hughes et al., 1996). In normal vision, we bring only a few points in the layout around us to the fovea, relying on the ambient system for the remainder. Yet, the Land Warrior device is one that uses focal visual information, but it has to be integrated with the operator's requirement for carrying out ambient visual activities. Peripheral vision also provides landmarks as to where some detail that was previously fixated (i.e., was clearly seen in central vision during a previous glance) lay relative to the feature presently being fixated, and these landmarks are likely to be unavailable within a small field of view. A viewer can compensate to some extent for a small field of view by making more head movements to sample the environment. But successive small glimpses of the environment that are obtained by such movements (which are much slower and more cumbersome than eye movements, especially with heavy helmet-mounted displays in place) can provide information about the entire object, or scene, only if they are effectively stitched together in memory, which is not necessarily possible in cluttered or unfamiliar settings.

There is currently no single accepted cognitive theory from which one can set the bounds of an individual glimpse. For example, are successive views of a display "directly" placed by the visual system into a single perceptual setting, without passing through any memory-like encoding process, just so long as there is enough structural overlap between the views? If so, which seems doubtful (Hochberg and Brooks, 1996), how much overlap must there be? Over how much delay? Over how many shifts of view? Regardless of theory, it is known that reduced fields of view reduce a viewer's ability to grasp things and to maneuver within the visual environment.

According to both a large body of research and common sense, objects and scenes can be identified more rapidly and more correctly when a viewer has been previously set or primed by those objects or by the categories to which they belong (Biederman, 1985; Bachman and Alik, 1976). The context in which an object appears, if it is an appropriate context, can serve much the same function (Biederman, 1981), but that depends on a field of view sufficient to provide that context. Reduced fields of view eliminate or reduce the context and thereby its facilitating effects.

There may be some minimum field of view below which a wearer will be unable to achieve a coherent grasp of the context, even by making the successive head movements discussed above; this is suggested by aperture-viewing studies and by examining motion picture use of "establishing shots" (Hochberg, 1986). At a narrow field of view, the facilitating effects of the context on object recognition may be lost, and the context is often necessary for accurate perception. Soldiers commonly are required to drop to the ground, roll rapidly, and survey their surroundings. It seems likely that the effects of narrow fields of view (and protruding eyewear) may require special training on such tasks and warnings about specific vulnerabilities.

When it is important for a viewer to have a ready grasp of where people and things are distributed within the visual environment (which must often be the case with infantry soldiers), the higher cost and weight of displays that go with wider fields of view may be unavoidable. Only controlled research under field (or field-like) conditions can inform that decision. In any case, because the sequence of visual queries (e.g., successive glances and head turnings) is elective—depending on a viewer's task, knowledge, and attention as much as on the information provided by the visual display at each step in the sequence—one must consider the situation that the viewer needs to grasp and the factors that affect such situation awareness.

TRAINING

As mentioned earlier, problems with the monocular display in the Apache helicopter were at least partially alleviated by training. There are at least three quite different areas in which specific training may help infantry soldiers use the helmet-mounted display, and the effectiveness of such training should be evaluated:

1. Performance of manipulations and locomotion under the offsets described should be practiced, and relearning of behavior during the aftereffects of protracted sessions should be pursued to familiarize the soldier with the existence and nature of the aftereffects.
2. To execute certain tasks, soldiers will have to substitute head motion parallax for binocular stereopsis in order to gain depth information in Zones 1 and 2. Similarly, they will have to substitute search through head movements for search through eye movements because of the reduced field of view. Training to criterion in several critical tasks similar to what must be done in the field (e.g., setting fuses, replacing pins in grenades, clearing weapon malfunction in Zone 1 and detecting approaching threats in Zones 2 and 3) may help decrease the costs of these informational losses.
3. Objects and terrain seen through these devices, especially narrow field of view thermal imaging, do not present the familiar perceptual units that so quickly and seamlessly serve to build our normal visual world. It is more like recognizing planes by radar signatures, but trying to do so in the course of rapid movement through a cluttered environment. Fortunately, practice with the purely visual task of recognition and identification can be obtained as much as is necessary using recorded and/or simulated displays. How effective such training is, and how much is needed, are questions for research.

CONCLUSIONS AND DESIGN GUIDELINES

The Land Warrior System aims to increase both the effectiveness and the survivability of infantry soldiers, using technologies that include portable computers, satellite navigation, light amplifying and thermal sensors, and both helmet-mounted and hand-held displays. Although such technologies can certainly enhance performance under certain conditions, they incur costs and risks as well. The pros and cons of each innovation must be considered in balance, to avoid net reductions in safety and capability. Evaluations must be informed by real or simulated research in the field; they should not be based solely on analyses of human abilities in laboratory situations.

Field research to test the effectiveness of this equipment has only recently begun. To be effective, the research must be directed toward conditions under which the net benefits from specific sensory enhancements are of questionable value. On the basis of the relevant research literature, this chapter summarizes what planners need to know in order to assess the benefits and costs of the major proposed enhancements. We believe that carefully designed field testing, guided by the kinds of human factors issues that are raised here, together with the concerns expressed by those who use the equipment, will be needed continually as this program evolves.

1. The proposal to use a monocular display appears to be motivated by the lower weight and cost of this configuration, as well as the desire to maintain dark adaptation in one eye. Our review has pointed out that the monocular display may result in rivalry, which can induce fatigue and disorientation. In addition, stereo depth information will be lost, which is an important depth cue when contrast is poor and obstacles are within 30 meters. Field tests tend to support this concern. We recommend that a binocular display be seriously considered and further field tests be conducted to evaluate the effects of display configuration on a variety of soldier tasks.

2. Our analysis suggests that a variety of depth cues are degraded by limited display resolution and field of view. This in turn should impact task performance within the three different depth zones of action. We recommend that field studies be conducted to determine how resolution and field of view affect performance in the three zones. In addition, training in making head movements and scanning patterns, which may partially alleviate these problems, should be investigated.

3. Thermal imagery presents a special challenge to the soldier's visual system because many of the usual cues to depth and shape available in visible light are absent in thermal images. Once again, training may be particularly important in the successful use of the thermal images.

4. The effects of long-term use of monocular displays are unknown. This issue should be investigated before a monocular configuration is adopted.

5. The use of the helmet-mounted display for maps and other symbology may be problematic. Symbology tends to produce clutter and may interfere with the perception of the sensor imagery. Maps and certain other kinds of symbology might be better displayed on a hand-held or wrist-mounted device.

6. The use of off-axis sensors, such as the image intensifier mounted on the helmet, may produce a variety of illusions, disorientation, and aftereffects. This placement should be avoided if at all possible.

7. Placement of additional weight on the helmet raises concern over fatigue, increased physical workload, and related increases in cognitive workload. The helmet-mounted display should be evaluated under the demanding physical conditions in which these interactions are likely to occur.

TECHNICAL NOTE: VISUAL ACUITY AND RESOLUTION

The preferred way to describe the minimum size target that can be seen is in terms of the visual angle it subtends at the viewer's eye in units such as arc minutes (Ogle, 1953). Many other units are also in use (see Figure 4-2). One measure is visual acuity, which is usually defined as the reciprocal of the target size in arc minutes (of subtended visual angle). One implication of the visual acuity unit is that normal vision corresponds to 1 arc minute (equivalent to Snellen acuity of 20/20).

In clinical practice it is common to use the Snellen fraction. The numerator of this fraction is usually taken as 20 and the denominator (usually in multiples of 10) is the range at which a young viewer with no visual abnormalities or dysfunction could discriminate alphanumeric characters that the testee can see at 20 feet (e.g., if your vision is 20/200, that means you need to be at a viewing distance of 20 feet to see letters a "normal" viewer could see at 200 feet, and you would be unable to read this text). While the measure has many limitations and acuity does *not* equal resolution, it is a commonly used reference.

Line resolution requirements (RCA, 1968): in television terminology, a line refers either to an actual scan line or to the time period allocated for a scan line. By this last definition, commercial broadcast TV in the United States is a 525-line system. Less than 525 actual scans are possible, however, because approximately 35 of the periods are used for the vertical retrace. Thus the number of actual or active TV lines is 490. Applying a Kell factor of 0.7 to this figure gives the equivalent of 343 active lines for use in considering resolution capabilities. (Because the phase relationships between a scanning spot and the objects in a natural scene cannot be controlled, some loss of resolution results. A commonly

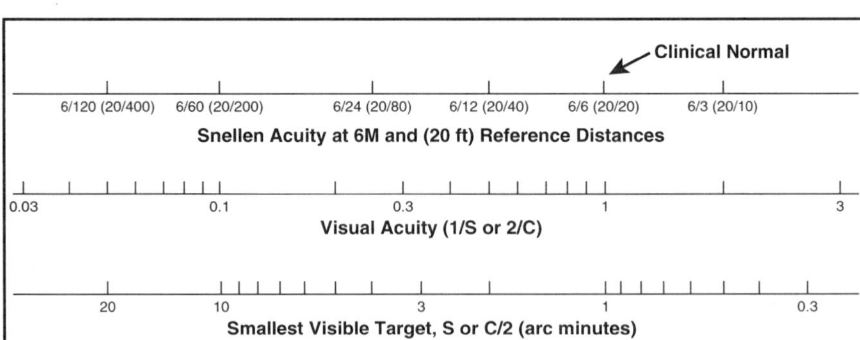

FIGURE 4-2 Visual acuity units. Source: Farrell and Booth, 1984. Reprinted by permission.

TABLE 4-4 Line Resolution Requirements

Task	Line Resolution Per Target Minimum Dimension
Detection	1.0 ± 0.25 line pairs
Orientation	1.4 ± 0.35 line pairs
Recognition	4.0 ± 0.8 line pairs
Identification	6.4 ± 1.5 line pairs

used figure is 30 percent. Thus the number of lines for effectively calculating resolution is 70 percent of the total. This value is known as the Kell factor.)

Furthermore, since an active TV line can represent, at most, one-half of a cycle of a periodic target (a light or a dark bar), at least two lines (a line pair also used as a measure in night vision goggles) are required to represent one cycle of a periodic target. It is important to keep this ratio of 2 active TV lines per cycle of spatial frequency in mind when dealing with line-scan systems (note that scan lines typically describe vertical resolution; horizontal resolution measures refer to pixels). This may be mitigated to some extent with helmet-mounted displays in which head movements can cause the sensor to move (i.e., scan line or pixel boundaries can be shifted) but data to demonstrate are not available, and for fixed displays (e.g., maps) the line pair/pixel pair requirement remains.

Angular threshold of the eye (RCA, 1968): the probability of seeing an object is influenced not only by the field luminance, the contrast of the object with respect to the scene background and the complexity of the scene, but also by the angular subtense of that object at the eye of the observer. Whereas under ideal conditions the eye can resolve down to 30 seconds of arc, the common figure used is 1 minute of arc. In most practical situations, however, the angular threshold of the eye is higher. With a high resolution complex image, for which line resolution does not enter as a limiting and confounding factor, it appears that 6 to 12 minutes of arc are required for typical visual acquisition and recognition tasks. Table 4-4 summarizes conclusions from one set of measurements of the capability of humans to perceive single military targets (standing man to tank size) as a function of the limiting resolution per target minimum dimension (Johnson, 1960).

Resolution example: an SVGA computer screen rated at 1,280 pixels × 1,024 lines, at a viewing distance that would result in a horizontal screen subtense of 10° horizontal by 7.5° vertical (with a 3/4 aspect ratio) could be defined as having resolution as follows:

horizontal: 1,280/10 = 128 pixels/degree of visual angle
or
128/60 = 2.13 pixels/arc minute of visual angle
or
1.07 pixel pairs/arc minute, which is approximately equivalent to 20/20 Snellen acuity.

In the case of the Land Warrior System, a 40° horizontal by 30° vertical field of view is subtended by a (nominal) 640 × 480 pixel display. Taking a 0.7 Kell factor into account, however, active lines/pixels are actually 448 × 336, resulting in a resolution of 5.35 arcmin/pixel or a useable resolution of 10.7 arcmin/pixel. A Snellen equivalent measure for acuity would be 20/214.

5

Auditory Factors in the Design of Displays and Controls

The ambient sound environment of the dismounted soldier is likely to be extremely varied. At one extreme, the background noise will be so loud as to preclude any sensible message transmission—either incoming or outgoing. Such noise can be a serious source of stress (see Chapter 6) as well as an interference to communication. At the other extreme is the need for surreptitious activity in a quiet environment wherein any audible sound generated by the soldier is to be avoided for security reasons.

Either of these ambient conditions can restrict the utility of auditory subsystems. Still, high visual channel loadings and priority occupation of hands means that auditory displays and controls may offer real advantages.

AUDITORY DISPLAYS

The visual channel is the mode of choice for providing information at high rates to the dismounted infantry soldier. However, in certain tasks and situations an auditory display may be more appropriate. Auditory displays are frequently used for alerting, warnings, and alarms—situations in which the information occurs randomly and requires immediate attention. The near omnidirectional character of auditory displays is a major advantage over other types of helmet-mounted displays. Table 5-1 summaries some of the factors to consider when making a choice between an auditory and a visual display.

The individual soldier's computer/radio is the main source of the auditory information for the Land Warrior System. It is currently envisioned that the auditory displays will be presented to the soldier via a headset mounted in the

TABLE 5-1 When to Use the Auditory Versus Visual Form of Presentation

Use auditory presentation if:	Use visual presentation if:
1. The message is simple.	1. The message is complex.
2. The message is short.	2. The message is long.
3. The message will not be referred to later.	3. The message will be referred to later.
4. The message deals with events in time.	4. The message deals with location in space.
5. The message calls for immediate action.	5. The message does not call for immediate action.
6. The visual system of the person is overburdened.	6. The auditory system of the person is overburdened.
7. The receiving location is too bright or dark adaptation integrity is necessary.	7. The receiving location is too noisy.
8. The person's job requires him or her to move about continually.	8. The person's job allows him or her to remain in one position.

Source: Deatherage (1972: Table 4-1).

helmet, although they also have the capability of interfacing to a handset. The auditory displays are currently envisioned to be either monaural with the handset or biaural[1] with the integrated headset. In this chapter we discuss the characteristics of auditory displays as well as some specific guidelines for their design.

Detectability of a Sound

An auditory signal is detected with increasing probability as the level of the sound increases. The masked threshold is defined as the level required for 75 percent correct detection of the signal when presented to a listener in a two-interval task. In a two-interval task, the listener reports which one of two noise intervals randomly contains the signal. Some guidelines on setting the level of an auditory display are based on Sorkin (1986):

1. A signal 6 to 10 dB above the masked threshold allows near-perfect detection in controlled conditions.
2. Signal levels 16 Db above the masked threshold will be sufficient for situations requiring a rapid response to a signal, such as a warning signal.
3. The level of an auditory warning signal should be less than 30 dB above the masked threshold, in order to minimize operator annoyance and the disruption of communications.

[1]Biaural presents the same signal in both ears. This is not the same as stereo presentation. With stereo presentation the signals not the same as both ears.

4. Nonauditory channels should be considered for environments that require sound levels above 115 dB.

Two factors that affect the determination of the masked threshold are the spectrum and duration of the interfering sound. More masking occurs when the signal and the interference are close in frequency, especially when the frequency of the interference is below that of the signal. As the intensity of the interference increases, the effect spreads to additional signal frequencies. When the signal is shorter than 100 msec, the masked threshold level depends on the signal energy rather than signal power and is therefore more difficult to detect than longer signals. In the operational environment of land warrior, the determination of a single masked threshold and therefore a single level for the auditory display may be difficult. The acoustic environment has a wide variety of interfering sounds that may or may not be present at any one time. The environment may range from complete quiet to the roar of battle, with both wide-band noise and impulse noise from weapons fire and explosions. An adaptive system that monitors the acoustic environment and sets the warning appropriately should be investigated.

Tonal Displays

A sound composed of the related components of 1,250, 1,500, 1,750, and 2,000 Hz has the same pitch as a single component of 250 Hz. This low-frequency pitch is perceived even though no signal energy is present at low frequency and even in the presence of interfering noise at low frequencies. This so-called missing fundamental frequency is heard because of the sensitivity of the auditory system to the harmonic structure of sounds. This harmonic pitch provides a useful code for auditory displays. The missing fundamental is stable and relatively insensitive to the relative levels of the component frequencies or to masking of some components, provided there is a sufficient number. Specific design criteria for tonal displays are informed by Patterson (1982):

1. The pitch of warning sounds should be between 150 and 1,000 Hz.
2. Signals should have at least four dominant frequency components, within the first 10 harmonics, in order to minimize masking effects, minimize pitch and quality changes during masking, and maximize the number of distinctive signals that can be generated.
3. Signals should have harmonic rather than inharmonic spectra.
   a. Lower-priority warning signals should have most of their energy in the first five harmonics.
   b. Higher-priority, immediate action signals should have more energy in harmonics 6 through 10.
   c. High-priority signals can be made distinctive by adding a small number of inharmonic components.

4. The frequency range of the signal should be restricted within 500 to 5,000 Hz, with the dominant frequency components within the range of 1,000 to 4,000 Hz.

Temporal form and shape of auditory displays are important factors for detectability, coding, and listener reaction. The following guidelines for this dimension of auditory displays are based on Patterson (1982):

1. Near-optimal envelope parameters are a minimum of 100 msec duration, a 25 msec rise and fall time, and quarter sine shaping.
2. Onset rates of less than 1dB/msec with the final level falling below 90 dB.
3. Use a variety of temporal patterns in order to minimize confusion.
4. Code urgency or priority with pulse rate (i.e., code high urgency with high pulse rate).

In a study of aircraft warning signals, Patterson and Milroy (1980) found that, although large sets of warnings can be learned, considerable learning time and regular retraining is required. Patterson's (1982) recommendations are that no more than six immediate action warning signals and two attention signals should be used, provided distinctive temporal and spectral patterns are used, the perceived urgency of the warnings matches their priority, and warning sounds are followed by keyword speech warnings. An attention is a special warning sound that signals the priority level of the following warning.

Speech Displays

For the dismounted infantry soldier, especially one with a helmet-mounted display, speech displays should be considered as a means to relieve the possible information overload of the visual channel. Deathridge (1972) gives the following reasons for using speech rather than other auditory signals in auditory displays:

1. Flexibility.
2. Ability to identify the message source.
3. Listeners don't have or need special training in coded signals.
4. Rapid, two-way exchange of information is required.
5. Messages deal with the preparation for a future event.
6. Situations of stress might cause the listener to forget the meaning of a coded message.

Issues that need to be addressed for speech displays are: (1) What are the optimal ways to generate speech displays? (2) What is the best way to integrate them with other displays and tasks? In this section we present some general principles and guidelines for speech displays, which are based largely on studies of cockpit displays but should be appropriate to the Land Warrior System.

Intelligibility is the most commonly used performance measure for speech displays and speech communications systems in general. Speech intelligibility is the percentage of utterances correctly recognized by the listener from a set of utterances presented under a given listening condition. Several intelligibility tests that have been developed and commonly used are the Modified Rhyme Test (House et al., 1965), the Diagnostic Rhyme Test (Voiers, 1977), and phonetically balanced words. Intelligibility measures can be used to determine how sensitive speech displays will be to disruption by noise.

Determining the appropriate level for speech displays is more involved than that of other auditory displays. Considerable speech information is carried in the consonant sounds. These sounds have shorter durations, higher frequencies, and lower power than vowels. In a high noise environment, the level of the vowel must be much higher than the background noise in order for the consonants to be detectable. In such situations, preprocessing the speech with either a 3dB/octave boost or peak clipping can improve the intelligibility of the speech in noise.

Synthetic speech systems allow considerable control of speech parameters such as pitch, speech rate, sex, and accent. This allows the generation of speech displays that have distinctive characteristics from speech heard over communications channels. One drawback to synthetic speech is it is much more sensitive to the effects of linguistic and task context, the level of operator training, background noise, and other manipulations of the spectrum. Pisoni (1982) has shown that synthetic speech may require more attentional resources than natural speech.

One ongoing debate in the design of speech displays is the message format and the use of auditory alerts preceding the speech message. The major debates relate to the relative effectiveness of monosyllabic versus polysyllabic words, keywords versus sentences, and speech messages with or without preceding alert tones.

For both natural and synthetic speech, polysyllabic words are more intelligible than monosyllabic words. Similarly, words in sentences are more intelligible than words in isolation. The context provided by the additional syllables in the words and/or the words in the sentence increases the redundancy and improves intelligibility.

Because of the ongoing use of the speech channel for other purposes and the difficulty of communicating over a long speech warning, Patterson (1982) advocates limiting the use of speech warnings for immediate-action emergency conditions. He suggests the use of sentence-length speech messages when signaling abnormal conditions that are less time critical. Other use of sentence messages are when disruptions are possible and the number of alternative messages is large. Patterson's integrated approach to auditory warnings is presented in Table 5-2.

Simpson and colleagues (1986), using a different design philosophy, recommend the use of speech only for the most time-critical warnings, the use of a

TABLE 5-2 Integrated Auditory Warning System

Priority	Purpose	Result	Description
Highest	Emergencies	Immediate	Tone sequences with key-word warning
Second	Abnormality	Immediate	Specific tone prefix awareness with voice message
Third	Advisory	Check visual display	Specific auditory signal

distinctive voice, no nonspeech alerting signals, repetition of the speech warning only after the operator has had enough time to correct the problem, and four-syllable minimum speech messages; the shorter 4 to 8 syllable messages present information more rapidly and cause less interference with other communication. In their view, the distinctive voice with a machine quality precludes the need for an alerting tone prefix. Patterson's philosophy is that the voice message is a backup to the aural signal and may provide advisory information. For different types of tasks, one or the other design philosophy may be more effective.

Table 5-3 can be used as a guide to select what type of functions are appropriate for the different types of auditory displays.

An additional capability with the use of speech displays is the ability to store messages for playback at a later time (i.e., voice mail). This would provide the soldier with the ability to save incoming voice messages when he is not able to attend to them or to save important messages he may want to review again. Some indication that messages are stored would be needed, like the message light on an answering machine. Not all soldiers may want or need this kind of auditory display. It may be more appropriate for platoon or squad leaders rather than individual squad members.

Three-Dimensional Auditory Displays

Spatial hearing technology has developed to the point at which it is now possible to present spatial information to a listener using headphones. The applications of three-dimensional auditory displays to the infantry environment include: indicating the location of other soldiers, threats, and targets; introducing spatial separation among communications channels; and providing an auditory beacon as a navigation aid.

Modern views of spatial hearing suggest that localization judgments depend on three classes of acoustic cues: interaural time differences, interaural level differences, and direction-specific frequency shaping of the high-frequency spec-

TABLE 5-3 Functional Evaluation of Audio Signals

	Types of Signal		
Function	Tones (Periodic)	Complex Sounds (Nonperiodic)	Speech
Quantitative indication	**POOR** Maximum of 5 to 6 toners absolutely recognizable.	**POOR** Interpolation between signals inaccurate.	**GOOD** Minimum time and error in obtaining exact value in terms compatible with response.
Qualitative indication	**POOR-TO-FAIR** Difficult to judge approximate value and direction of deviation from null setting unless presented in close temporal sequence.	**POOR** Difficult to judge approximate deviation from desired value.	**GOOD** Information concerning displacement, direction, and rate presented in form compatible with required response.
Status indication	**GOOD** Start a stop timing. Continuous information where rate of change of input is low.	**GOOD** Especially suitable for irregularly occurring signals (e.g., alarm signals).	**POOR** Inefficient; more easily masked; problem of repeatability.
Tracking	**FAIR** Null position easily monitored; problem of signal-response compatibility.	**POOR** Required qualitative indications difficult to provide.	**GOOD** Meaning intrinsic in signal.
General	Good for automatic communication of limited information. Meaning must be learned. Easily generated.	Some sounds available with common meaning (e.g., fir bell). Easily generated.	Most effective for rapid (but not automatic) communication of complex, multidimensional information. Meaning intrinsic in signal and context when standardized. Minimum of new learning required.

tra introduced by the head and pinnae. Kistler and Wightman (1992) argue that listeners determine the laterality of sounds based on interaural time differences and interaural level differences and distinguish between front and back and between up and down on the basis of spectral cues. Wightman and Kistler (1992) showed that low-frequency interaural time differences are the dominant cue for sound localization. Therefore, for signals to be localized easily, they should include frequency components that spread across the entire spectrum.

This sensitivity to interaural time and intensity also enables the human listener to detect signals that otherwise would be buried in the noise. One can demonstrate as much as a 15 dB advantage in detection when there are interaural differences in the signal and noise input to each ear. That is, the level of the signal can drop 15 dB below the level detected with identical inputs to both ears or with monaural listening.

In addition to improving detectability, these cues also facilitate improved processing of speech messages in noise. Speech intelligibility studies by Bronkhorst and Plomp (1988, 1992) have shown a 6 to 10 dB advantage with speech at 0 degrees azimuth and noise off-axis compared with the control condition of speech and noise at 0 degrees azimuth.

Spatial hearing also enables us to selectively attend to spatially separated conversations in a crowed, noisy room—the so-called cocktail party effect. (In such an environment, covering one ear tightly will cause a sudden decrease in the ability to hear separate conversations.) Ericson and McKinley (1996) studied speech intelligibility when two competing messages were presented spatially separated via headphones. Results showed a 10 to 40 percentage point improvement in intelligibility when compared with the control condition of both messages presented 0 degrees azimuth. Interestingly the greatest improvement occurred for messages in which both competing talkers were female and the least improvement when both talkers were male.

To support the use of three-dimensional audio displays for the infantry soldier, two additional technologies are required: stereo headphones and head trackers. In order to minimize the acoustic isolation, the stereo headphones should provide a minimum of attenuation to external signals (i.e., they should be acoustically transparent). The ear-rest-type earphone of the proposed Land Warrior System are not stereo and may cause significant attenuation and discomfort when inserted deep into the ear channel to maintain stability during the strenuous maneuvers the soldier will experience.

In another related concept that has been proposed, audio speakers are suspended inside the helmet shell. This arrangement would provide no interference with unaided hearing. The disadvantage is acoustic transmission of the auditory signal to the environment; this would be a significant disadvantage during covert operations. The arrangement would also cause problems for three-dimensional audio displays, since current systems require circumaural stereo headphones or ear insert devices. At least one manufacturer of 3DAD systems makes an acous-

TABLE 5-4 Advantages and Disadvantages of Monaural, Biaural, Stereo, and 3DADs

	Advantage	Disadvantage
Monaural	Inexpensive Single earphone	Sounds can not be localized Reduced signal detection Reduced speech intelligibility
Biaural	Small increase in signal detection	Headphones required
Stereo	Increased signal detection Increased speech intelligibility Minimal increase in complexity	Headphone required
3DAD	Improved signal detection Improved speech intelligibility Signals localizable Multiple channels monitoring Waypoint navigation	Increased complexity Headphones required

tically transparent headset with which environmental signals such as speech can be easily heard. Whether this technology would provide the same capability as unaided hearing is an area that should be investigated. Also, the ability of such a headset technology to operate in a rugged military environment is still to be determined.

A means of determining the soldier's head position and orientation is required to most effectively utilize three-dimensional audio displays. This provides the capability to fix the auditory signal in space as the soldier rotates his head. This capability may also be of value for some types of visual information displays as well. The most common forms of head trackers use magnetic or ultrasonic sensing. Head trackers of this form are commonly used in virtual reality and cockpit applications in which there is a fixed range over which the transmitter and receiver can operate. New technologies being developed use miniature gyroscope or magnetic earth sensors. Table 5-4 summarizes the advantages and disadvantages of monaural, stereo, and three-dimensional audio displays.

When combined with the global positioning system, the three-dimensional audio display could present the radio communications of an individual's squad members in the direction they are relative to him. This would improve situation awareness and could reduce fratricide. This would be advantageous for the desired capability called conversation mode communications, which is defined as the capability for three or more stations on the radio network to communicate simultaneously with each other. Another application would be way point navigation. A tone could be presented in the desired direction of movement; the soldier would follow the tone until he reached the desired way point; at that time, the tone would either cease, indicating the desired position was reached, or another tone would be displayed indicating the new direction to follow. The same technique could be used to designate targets for directing fire. Each of these applications is being investigated by the Air Force for applications in fighter cockpits. For the Land Warrior System, not all soldiers may need this capability. For example, platoon or squad leaders could have the capability to locate his squad members, whereas special operations forces or scouts would need the navigation capability during night operations or adverse weather.

AUDITORY CONTROLS

In a broad sense, an auditory control is a machine activated by any sound. However, for the helmet-mounted display we focus our discussion on voice commands from the user and the attendant need for voice recognition capabilities. For the dismounted infantry soldier, speech recognition can be of considerable benefit in concurrent task situations: (1) when both hands are engaged (such as soldier aiming a weapon); (2) when the eyes are engrossed in visual processing and cannot easily change gaze for manual data entry (e.g., forward fire control); and (3) when even only a single hand is engaged in a manual task (e.g., squad leader deploying troops).

It has been reported that speech recognition is as good or better than keyboard data entry, although error rates may be higher depending on the specific system, vocabulary, and task situation (error rates as low as 1 percent have been reported). However, speech recognition is not without its drawbacks. Feedback from a failure to correctly recognize an utterance may disrupt the concurrent activities, a disruption that speech recognition was intended to eliminate. Moreover, the same conditions that limit auditory displays—that is, either loud ambient noise conditions or surreptitious operations will strictly limit what and how the soldier can articulate voice commands. Since the circuitry needed to allow machine recognition has some cost and adds weight to the soldier's pack, the cost-benefit determination is uncertain. A step toward reducing that uncertainty can be provided by a brief exploration of the technology for voice recognition subsystems.

Types of Speech Recognition Systems

Speech recognition systems are classified along several dimensions: (1) the number of speakers they recognize, (2) the type of speech they recognize, and (3) the size of the vocabulary. The following paragraphs will address each of these dimensions and some associated trade-offs in turn.

Speaker-dependent systems recognize speech from only one speaker. Speaker-independent systems recognize speech from many speakers. The performance of speaker-dependent systems is generally better than that of speaker-independent systems. Speaker-adaptive systems start out with speaker-independent templates and adapt those templates to the current speaker and therefore approach performance levels of speaker-independent systems.

The next dimension is based on how the speech recognition system handles word boundaries. Isolated-word recognition systems require a pause of 100 to 250 msec or greater between words. Connected-word recognition systems require a very short pause between words. Continuous-speech recognition systems require no pause between words and accept fluent speech.

The last dimension is based on the number of words the system can recognize. Vocabulary size is generally classified as small (less than 200 words), large (1,000 to 5,000 words), very large (5,000 words or greater) and unlimited (greater than 64,000 words). In general the computational resources required to perform speech recognition increase across each of these dimensions. For example, a speaker-dependent, isolated-word, small-vocabulary system would require the least computational resources.

Word recognition accuracy is the most commonly used measure of performance for speech recognition algorithms. It is simply the percentage of words correctly recognized. Errors can be assigned to one of three categories: (1) substitution error (one word is recognized as another); (2) insertion error (a word is recognized that was not spoken); and (3) deletion error (a word spoken was not detected). An analysis of the errors provides feedback on the vocabulary design. For example, consistent substitution errors may imply that those words are easily confused and the vocabulary should be modified. In general, the word recognition accuracy decreases across each of the above dimensions. For example, as the vocabulary size increases, the word recognition rate decreases.

Environmental Issues

Most applications of speech recognition have been in relatively benign environments such as office or laboratory settings where the background noise is low. Speaker-dependent isolated-word and connected-word systems with 30 to 150 word vocabularies can provide greater than 95 percent accuracy in cockpit environments with high noise. Whereas previous work has characterized the acoustic environment for military weapons systems such as tanks, helicopters, and aircraft and examined the effects on speech recognition performance, little such work has been done for the battlefield. For the dismounted infantry soldier, other environmental factors are also present that will change the way he speaks and could negatively impact speech recognition performance. Things such as physical and mental fatigue, sleep deprivation, and physical exertion are all areas that have had little or no systematic study of their effects on speech recognition performance.

Human Factors Issues

Several human factors issues involved in the design of a speech recognition system are (1) vocabulary size, (2) vocabulary selection and syntax, (3) operator training, and (4) system training. Vocabulary size is determined by both computational and storage constraints on one hand and the required task on the other. Large vocabularies often lower the accuracy of a system because there is a greater possibility of error. The vocabulary should be designed so that it does not contain easily confused words. The vocabulary should be chosen so that it uses terminology that is familiar and common to the user in that task.

The vocabulary syntax can be designed to provide a greater effective vocabulary size by enabling a subvocabulary at each point in the command sequence. This also enables high recognition accuracy because there is a smaller number of words to choose from at each point. The syntax should also be chosen so that it is familiar to the user performing the task.

For isolated word systems, the operator must learn to insert a pause between each word. For speaker-dependent systems, the system itself must be trained for each speaker with 4 to 10 repetitions of each word in the vocabulary. Thus, for large vocabularies, hundreds of words must be spoken. This sort of training is very tedious and puts practical limits on the vocabulary size.

CONCLUSIONS AND DESIGN GUIDELINES

When the visual channel is heavily loaded, as is a distinct possibility in the Land Warrior System, the auditory channel may be an appropriate alternative way to convey information to the infantry soldier. The auditory channel has naturally evolved to serve a warning or alerting function. Issues to be addressed when using auditory displays are masking and interference by other signals, confusability between signals, training requirements, and signal localization. The impact on unaided hearing depends on the type of auditory display chosen. If monaural systems are chosen, there will be no interference with unaided hearing. A more problematic case is represented by use of circumaural stereo headphones. Both laboratory and field experiments need to be conducted to determine the impact on overall task performance for the types of being considered for use in the helmet-mounted display.

An overall issue to be addressed when using multiple displays is what information should be placed on what display and when. Are there certain tasks or stress situations when this allocation should change? (See Chapter 6 for additional discussion on these questions.) Little research has been conducted looking at the relationship between different display modalities (i.e., auditory and visual), the physical environment, and workload. Understanding this relationship and the trade-offs involved is critical for the successful use of such a complex system as Land Warrior System, which will be used in a highly dynamic and wide-ranging set of environmental and workload conditions. This section has presented guidelines for the development and use of auditory displays. Specific studies for the Land Warrior System need to be performed to validate the guidelines for the particular conditions of use.

On the output side, speech recognition might provide a means of control and information input not presently available for the many hands-busy, eyes-busy tasks in which the dismounted infantry soldier is engaged. Although speech recognition technology has made significant strides in the past several years toward large-vocabulary, speaker-independent, continuous speech systems, these advancements may not be practical in the near term for the dismounted soldier due to the computational resources required. Each type of speech recognition system has unique human factors and technology issues that will influence the final system design and performance. Trade-offs between the available computational resources and task requirements will need to be studied.

6

Stress and Cognitive Workload

The integration of very high-tech equipment into standard operations is a radical change in the challenges faced by the infantry soldier. In addition, the battlefield, in all its forms, remains a place of extreme stress. Coupled with this stress comes the new burden of cognitive workload associated with the operation and management of new technological systems. The Land Warrior System, as currently conceived, is but one version of a potential family of advanced systems—each of which may generate its own combination of stresses. In this chapter we examine these stress-inducing factors, to identify sources of potential problems and to recommend avenues to solve such problems, either through existing capabilities or by proposing additional research.

A UNIFYING FRAMEWORK

One of the reasons it is difficult to predict accurately the performance of soldiers under stress is that the scientific foundation of this area is inadequate. For many decades, we have relied on the untestable, inverted-U theory of physiological arousal, with both high and low levels of arousal inhibiting attention (see Yerkes and Dodson, 1908). This theory has been used to account for a wide pattern of performance, after it has happened; unfortunately, it has little if anything to say about what will happen in the future under specific circumstances (Hancock, 1987). Consequently, although the theory is very useful to the scientist trying to explain a confusing pattern of results, it is of little use to the designer of equipment or the leader of a platoon trying to predict how individuals will react to specific conditions. More recently, researchers have looked at effects on

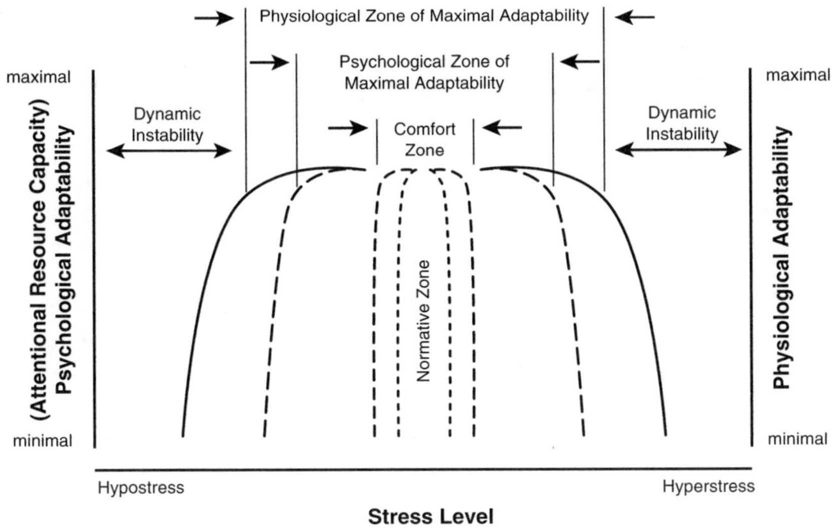

FIGURE 6-1 A model for the prediction of stress effects.

either a task-by-task basis (e.g., memory versus decision-making) or on a stress-by-stress basis (e.g., heat versus noise) (Hockey and Hamilton, 1983). However, these approaches provide no unified basis for understanding stress and subsequent effects on workload.

A unified approach to the prediction of stress effects has been presented recently by Hancock and Warm (1989). The basis of this model is shown in Figure 6-1. Briefly, the model establishes ranges of adaptability. Violation of these ranges of adaptability causes progressive failure in performance. Because the model uses both a physiological and a behavioral index, it can deal with the question of combined physical and mental demands and is recommended for use in association with the testing of advanced technologies. In this work, Hancock and Warm (1989) have combined existing physiological and psychological theories of stress effects into a unified view.

In the center of the figure is a region in which stress exerts little impact in performance variation. In the center of this comfort zone, individuals can be expected to produce their best performance. As stress increases, either through a systematic increase or a systematic decrease of appropriate stimulation, the individual is required to use greater resources to combat the influence of such demanding conditions. This reduces the resources available for explicit task performance, and so efficiency is reduced accordingly. It must be emphasized that the demands of the primary task itself may well act in this depleting fashion.

There are some strategies by which the individual can sustain satisfactory performance, even as stress increases. One common tactic is to select only those cues relevant to task completion and to reject of filter out cues that are irrelevant. This allows some degree of protection from stress effects. However, if the stress should continue to grow or should be perpetuated over a longer time, then some degree of performance failure is observed.

Individuals can tolerate a certain level of stress with no disturbance to their capability. This is the flat portion of the extended-U of the model. However, if the stress persists for a sufficient period or the severity is increased, the individual begins to lose capability. Since the functions stay the same, the way in which task performance falls off mirrors the way in which physiological capacity is impaired. The Army is accustomed to providing physiological support for its personnel, in terms of energy, resources, and specific equipment to avoid threat. The Army is accustomed to providing supplementary support when such systems fail or induce stress. The proposed model argues that new equipment ensembles and associated mission stresses can be ameliorated in the same fashion—that is, by providing training and supplementary support.

The model does not indicate where any individual will fail in relation to a specific form or level of stress. Furthermore, the model does not stipulate what stress alleviation or stress generation is inherent in the employment of the Land Warrior System. There is much to be learned about individual reactions under such circumstances, and this consequently represents one area of needed research. One critical aspect of this research must be the identification of incipient failure and the reduction of stress under the specific conditions associated with using the Land Warrior equipment. As is evident from the model, failure in performance, such as errors, will be evident before failure in physiological functioning. However, since the soldier is exposed to stress in order to accomplish a task, it is essential for his squad or platoon leader to be aware of these incipient failures. This implies the need for on-line feedback of individual soldier performance to be provided throughout the information network; it would then become possible for the leader to know when specific members of their command are becoming overloaded and subject to performance error and potential failure. Such information is critical to ensure mission success, and its transmission could be a central feature of the proposed helmet-mounted display.

Much more research effort needs to be directed to the critical question of performance prediction under stress. It is a major requirement for the success of the Land Warrior System on the battlefield of the 21st century. The Land Warrior System will add new tasks for the soldier to perform. These tasks are expected to be more information-intensive: they will require more reading and more cognitive processing than is currently necessary. The following section describes the existing sources of battlefield stress that provide the psychological and physical context for the introduction of the proposed system. These context factors must be considered in examining the implications of the helmet-mounted display for increased workload and, in turn, for soldier performance.

SOURCES OF OPERATIONAL STRESS

The Task as a Source of Stress

The traditional approach to understanding stress has been to focus on a specific source—for example, fatigue—and to try to understand how variation in fatigue affects capability. This view places primacy on the external source of stress. More recent views (Hancock and Warm, 1989) have emphasized the primacy of the performance task itself as the major source of stress. For the infantry soldier, mission requirements themselves have to be considered the proximal source of stress, and other outside influences, such as heat, noise, fatigue, etc., are seen as forms of stress that interact with the primary demands of the task. This new perspective is particularly important when we consider the cognitive demands associated with advanced technologies. New tasks initiated by the use of technologies, such as the helmet-mounted display, represent a major challenge and therefore a source of stress on the infantry soldier. In sum, the interaction among stressors may be crucial.

Acclimating to a source of stress provides resistance to its deleterious effects (Hancock, 1986c). For example, heat acclimatization many take a matter of some weeks but, having passed through such acclimatization, the individual is much better able to withstand extended exposure. This process of acclimatization is well known in physiological research. However, there is a comparable effect for cognitive systems. That is, to train high-performance skills (Schneider, 1985), it is necessary to engage in extended practice on the task—in essence, acclimating the cognitive (central nervous) system. To accomplish this, the task has to be constructed in a specific fashion. In particular, it has to be consistently mapped, between input stimulus and output response (see Schneider and Shiffrin, 1977). Consistently mapped, overlearned, or automated tasks are not only performed more quickly and with lower error rates, but also prove less vulnerable to the effects of external sources of stress (see Hancock, 1984, 1986a). As a consequence, how cognitive tasks are structured has a direct influence on how well they will be performed on the stress-replete battlefields of the future in which the Land Warrior or a comparable system might be employed (see Schneider, 1985, for further details on structuring complex performance tasks for optimal training). In the past, the military has led the way in terms of advanced training and it is especially certain that training will play a significant role in operating new technical support systems such as the Land Warrior System.

Information Overload and Underload

Many pursuits can be characterized as "hours of boredom and moments of terror," including aviation, law enforcement, firefighting, and even less life-threatening ones such as acting. Definitely included with these are the activities of infantry soldiers, who are often held at some location for many hours waiting, only to find themselves in the midst of a confusing engagement a short time later. In these circumstances, the individual goes from the potential extremes of underload to a massive overload in moments. Little is known about the effects of such transitions (but see Huey and Wickens, 1993).

What is clear is that, as a design principle, technology should provide soldiers with a compatible information load at all times. In essence, tasks should be structured as much as possible to transfer the information overload of engagement conditions to either the pre- or postengagement period. As much as possible, technical support should anticipate the demands of engagement and permit the preprocessing of pertinent information. Similarly, during the engagement, technology should off-load the information demands on the soldier, especially if noncritical tasks can be postponed for the postengagement period. In related work, Hancock and Chignell (1987, 1988) have suggested an adaptive human-machine system, in which the task of load distribution is subsumed by an intelligent interface (Chignell and Hancock, 1989), which uses signals from both the technology and the operator to decide which task components need to be given to which of the human-machine partnership and at what time. Such a system is under exploration for use in advanced single-seat tactical aircraft; in principle, it should be equally appropriate for the Land Warrior System.

Recent engagements have referred to the strategy, "Blind it, cut it off, and kill it." This refers to the process of destroying the enemy's information collection systems (sensors, etc.), the subsequent destruction of communication channels, and finally the destruction of an impotent enemy force. However, it is important to be aware that an information overload for our forces is just as dangerous as an information underload for the enemy. In depriving the enemy of their sources of information, our forces must avoid doing the same to themselves by producing an incomprehensible avalanche of information. This is especially true for the dismounted infantry soldier, who needs only a few sources of vital information during engagement. If this filtering process is not successful, there are liable to be additional cases of soldiers immobilized by stress—not the stress of battle but the stress of information overload. The use of electronic displays in the Land Warrior System opens the door wide to such a condition.

Information and Disinformation

Orthogonal to the problem of overload and underload is that of information versus disinformation. Intimately linked to the question of security, the problem of trust in the technical support system is an extremely important one. Lee and Moray (1992) found that the level of trust exhibited by the operator is a critical component in a system's utility. Riley (1994) found that operators are very slow to recover belief in system information, given a single failure event in an automated system. Given military experiences with friendly fire incidents, it is even more critical that the information supplied by an information system does not introduce distortions and inaccuracies. Of course, access to the force's information systems by the enemy forces represents a devastating event which could have negative influences even beyond a specific incident. Ideally, critical information would always be confirmed by a secondary source, preferably one not relying on technological support.

Physical Work

A great many occupations make taxing mental demands and a large (but diminishing) group of occupations impose physical demands. The job of the infantry soldier requires considerable physical work at the same time as critical decision making. As technical innovations percolate more into the soldier's everyday activities, these combined physical and mental demands are becoming a critical concern. In the panel's view a considerable research effort needs to be directed to an understanding of these combined influences.

In addition to the combined stresses associated with physical and mental workload, there is the stress generated by the equipment itself. It is important to keep in mind that the weight of the Land Warrior System that has to be carried is only one component of an already large assembly currently carried. Extreme stresses may become associated with operation of the equipment itself, especially helmet-mounted displays, if they are not comfortable and clear, and if the display resolution is poor and they are trying to operate vital controls. How the proposed equipment is integrated with NBC protection is a vital question when considering reducing sources of soldier stress. Experience indicates that equipment that is not considered vital or easy to operate during a firefight will be quickly jettisoned. Consequently, the predicted advantages of having well-equipped soldiers on the electronic battlefield may well be negated by the problem of discomfort. Consequently, considerable research is needed into the questions of comfort, prolonged use, systems integration, ease of ingress and egress, and fundamental ergonomic questions whose solution can reduce the stress associated with suboptimal design.

The Operational Environment

There is no greater form of stress than an immediate threat to life. Consequently, the proposed Land Warrior System will have to operate under the most extreme conditions of stress. Also, because there is no certain way to predict in what theater of operations soldiers will be required to perform, we consider specific forms of stress as well as the general, pervasive forms.

Stress accompanies military operations. Typically, stress is seen as a threat or an overload of demand conditions that seem likely to overwhelm the individual. However, stress is also destructive in the case of chronic underload, especially when accompanied by the uncertainty that marks contemporary operations. Provision of on-line, momentary information can therefore have a profound influence on experienced stress, especially when some form of control or autonomy is provided to the individual soldier.

Perhaps the most influential effect of stress in the present context is the restriction in the range of cue utilization. Attentional narrowing (which is highly related to the concept of situation awareness) occurs under extreme conditions. Easterbrook (1959) proposed, and others have in general confirmed, that the nature of the cues recognized in a display varies as a function of stress. This phenomenon is familiar to pilots, whose scan pattern alters directly with the threat of environmental conditions. Originally, this was theorized to be a visual phenomenon, something akin to G-induced loss of consciousness, in which the pilot proceeds from tunnel vision through gray out to black out and finally unconsciousness. However, more recent findings suggest that it is an attentional phenomenon, since by manipulation of cue salience "narrowing" can be directed to the visual periphery. Given the reality of this phenomenon, the design of portable equipment and the way in which information is provided must vary according to context (Wickens et al., 1993).

Physical Sources of Stress

Heat

Heat is one of the more ubiquitous and disruptive forms of environmental stress that is liable to be encountered. Recent operations of U.S. forces, including the Desert Storm, Grenada, Haiti, and Arabian Gulf missions, have involved the problem of heat stress. Not only is heat stress a consequence of the geographical location of operations, but also, in any encounter in which the use of chemical or biological weapons is suspected, heat stress can affect military personnel wearing exclusion garments for protection. The heat load on the operator, generated by these exclusion garments is considerable, and on occasion has been sufficient to cause missions to be canceled or mission goals to be changed. Heat stress from such use does not have to occur in the hot regions of the world.

Researchers typically focus on the complex effects of heat on central nervous system functioning. This orientation is understandable, given that heat stress investigations are often part of a more general for the effects of stress on human performance. However, in the present context, there are certain coarse-grained, physical disruptions from heat that have to be considered first. People under heat stress sweat, and sweat is a problem for any visual display, since profuse sweating can impair visual capabilities. Also, individuals brushing sweat from their eyes may miss information solely from this physical gesture and not from some more complex effect on cognitive activity. In fact, to make accurate statements about heat effects on cognitive function, one must be careful to exclude these peripheral effects.

Wearing a helmet-mounted display exacerbates these problems. A large percentage of heat load from the body is lost from the head, particularly from the front of the face. Covering this area is liable to generate profound heat-related problems. Heat loss also comes from vapor exchange in breathing. Displays that cover the whole face can lead to extensive fogging problems, as experienced for example in space suit operations. Although leaving half the face uncovered might seem like a solution, it ignores mandates about full exclusion garments for toxicological warfare. Sweat also presents a problem with respect to fit. Cooling off by injecting water into the interface between the display unit and the soldier's face can cause significant short-term and long-term problems, those of slippage and maintenance of complex electrical equipment exposed to water on a regular basis. Independent of all the concerns of geographical heat stress and garment heat stress, prolonged covering of the eyes causes thermal balance problems. Of necessity, any proposed displays need extensive field testing and also systems integration with existing equipment ensembles.

Predicting human tolerance to heat stress when physical load is expected can be derived from Figure 6-2. Predicting the breakdown in behavioral decision making from heat stress can be derived from Figure 6-3; this figure also includes information about body temperature, so it can evaluate the combined effects of physical and mental demand.

Cold

Whereas heat presents a number of unique problems, cold also provides considerable contextual challenge for advanced systems. Paradoxically, many of the heat-related effects, such as sweating and fogging, apply in the cold as well, since individuals are covered in heavy clothing to resist cold effects and are usually overclothed to provide protection against extended exposure. The primary, unique factor in cold exposure is shivering. Shivering is a physiological process designed to spend 100 percent of the body's stored energy to heat it. The oscillating motion experienced is not directed to behavioral goals and in fact directly interferes with them. Hence, the central problem of shivering is mechanical interference. Traditionally, the research literature has looked at how this interference debilitates psychomotor performance such as pursuit tracking; however, there are obvious effects on perception as well. Although fewer studies have examined this facet of performance, it is clear that some degradation is to be expected.

Unfortunately, by the time shivering becomes a mechanical problem, it has already begun to affect central functions such as speed of response and decision making. As a consequence, cold effects become as much of a contextual challenge as heat effects, although somewhat less likely to be encountered.

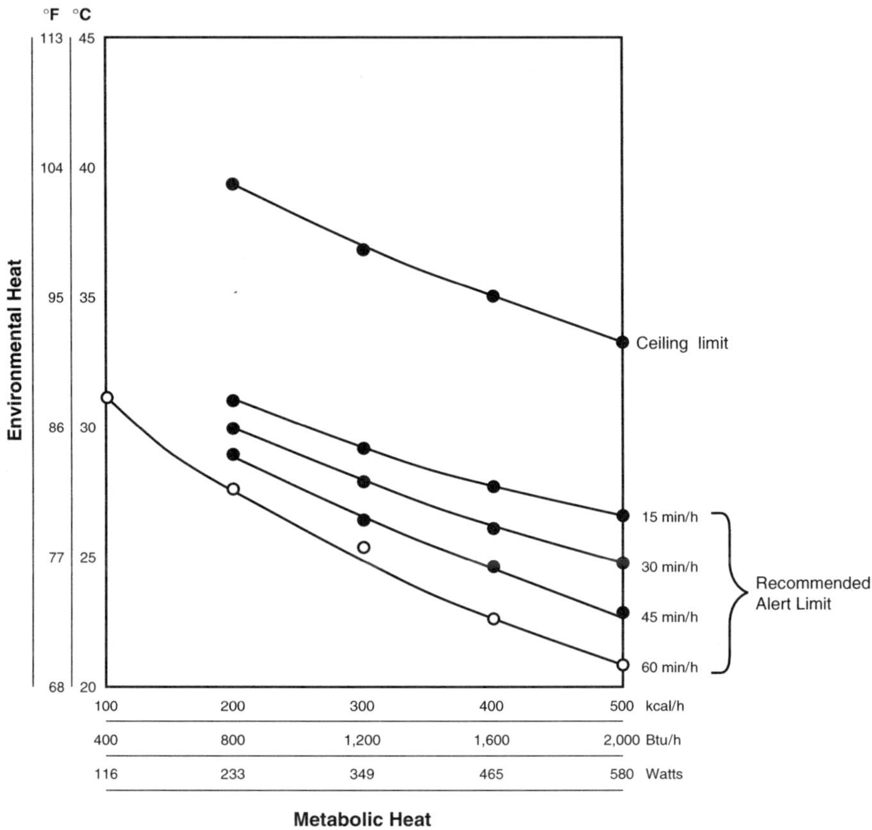

FIGURE 6-2 Recommended heat-stress alert limits for heat-unacclimatized workers. Source: National Institute for Occupational Safety and Health.

Noise

Noise is another form of disruptive stress, and its concurrent and aftereffects are less well understood than those of temperature. Noise effects are much more diverse and difficult to localize than other stresses. Unless noise drops below approximately 30 Hz, the effects are mainly aural, although very low frequency noise and vibration effects are directly linked.

Noise is an intermittent stress. In World War I, noise was one of the major forms of annoyance for the infantry soldier; partial deafness from shelling was not uncommon. Modern warfare relies less on such artillery tactics, since the theater of war itself is more mobile, so the individual soldier is more likely to experience brief bouts of intensive noise punctuated by bursts from ordnance impact. These may be interspersed with periods of relative quiet.

One obvious concern in physical ergonomics is the provision of auditory

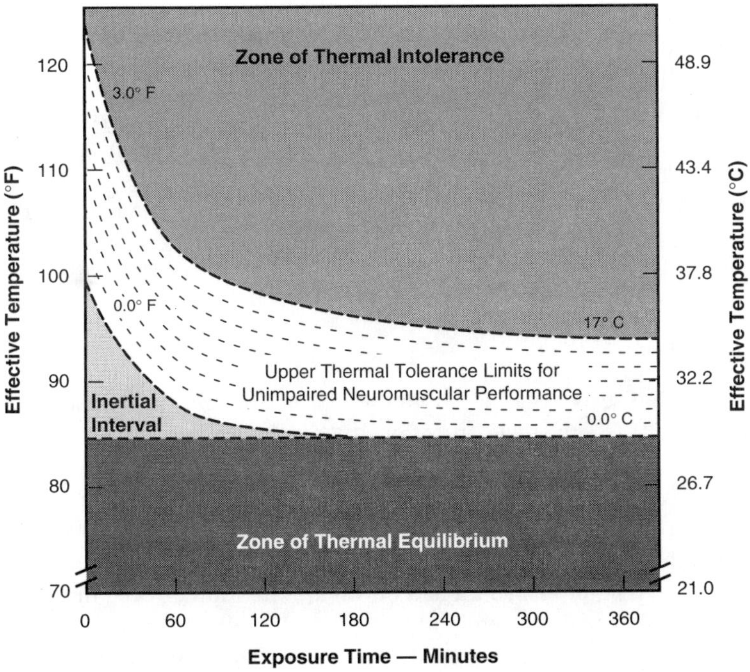

FIGURE 6-3 Recommended heat limits for cognitive performance. Using the axes of exposure time and stress intensity, three major zones of worker performance efficiency are distinguished. The heat stress axis is expressed in terms of effective temperature (ET). The zone of thermal intolerance describes a region of complete performance cessation due to physiological failure. The zone of thermal equilibration describes a ceiling level of ambient conditions that are insufficiently severe to disturb deep body temperature and thus curtail performance efficiency. Embedded between these zones are isodecrement contours, which describe the upper thermal tolerance limits for unimpaired cognitive and neuromuscular performance. Precise details for differing task characteristics are described in the text. The boundaries and contents of each zone can be described in terms of dynamic change to the core body temperature of the exposed individual that permits their use for occupations where traditional indices cannot accurately assess the heat load imposed on the worker. The inertial interval, which reflects the resistance of human core temperature to sudden change, completes the present picture. Source: Adapted from Hancock and Vercruyssen (1988).

messages against a noisy background. In a literal sense, this is a signal-to-noise problem. When soldiers are exposed to extremely noisy conditions, multimodal forms of information transmission may make sense, although that strategy may risk visual overload.

Noise is defined as unwanted sound, such as loud sounds that are either momentarily disturbing or more prolonged and annoying. Unwanted sounds could also include an excess of verbal communications and messages, especially if a party-line system is in use. The soldier may not want to listen to extraneous information that distracts him from the task at hand.

In any battle, noise and its persistent aftereffects will be present, and these factors must be considered when integrating developed technologies. It may be possible to develop a helmet-mounted system specifically designed to insulate the soldier from excessive ambient noise; however, sometimes vital auditory cues are needed to survive. Flexibility and choice of configuration—that is, an adaptive system—is the best recommendation.

Vibration

Whereas noise is airborne, vibration can be thought of as material borne. Vibration results from vehicle operations such as helicopter and armored personnel vehicles, and also as a function of locomotion. Vibration effects are very similar to shivering, with one critical exception. The frequency of material-borne vibration can vary over a considerable range, including the resonance frequency (3-6 Hz) of the major organs of the body. Vibration in this range is exceptionally disruptive to the point of nausea and beyond. Vibration effects, like shivering, are liable to affect both perception and action. Usually, vibration is masked by internal compensatory systems; we do not see the world bounce up and down as we walk. However, with helmet-mounted displays, the suppression of this intrinsic form of vibration is not clear. Much empirical research still needs to be done on work performance while walking (see Sampson, 1993). The panel urges that such research continue, since it lies at the heart of the interface between the physical characteristics of the environment and the design of the technical support systems themselves. Figure 6-4 illustrates the quantitative limits on vibration for visual capability. Boff and Lincoln (1988) discuss a wide range of vibration effects across differing sources of stress.

Extended Operations/Time Of Day

As mentioned earlier, infantry operations are conducted 24 hours per day. An important factor that remains uncertain is the length of an individual engagement and consequently the duration of mission that the infantry soldier is expected to sustain. In ground-based warfare of the immediate past decades, emphasis has been given to the first 72 hours of engagement. In consequence, the U.S. Army became a world leader in understanding the effects of sustained operations. Given the depth and clarity of this work, it would be redundant to repeat it here. However, it is important for any designer and user of the Land Warrior System to consider the stresses associated with circadian phases of operation, duration of operation, and associated fatigue. Whereas the Army has an outstanding record in the area of extended operations research, specific studies

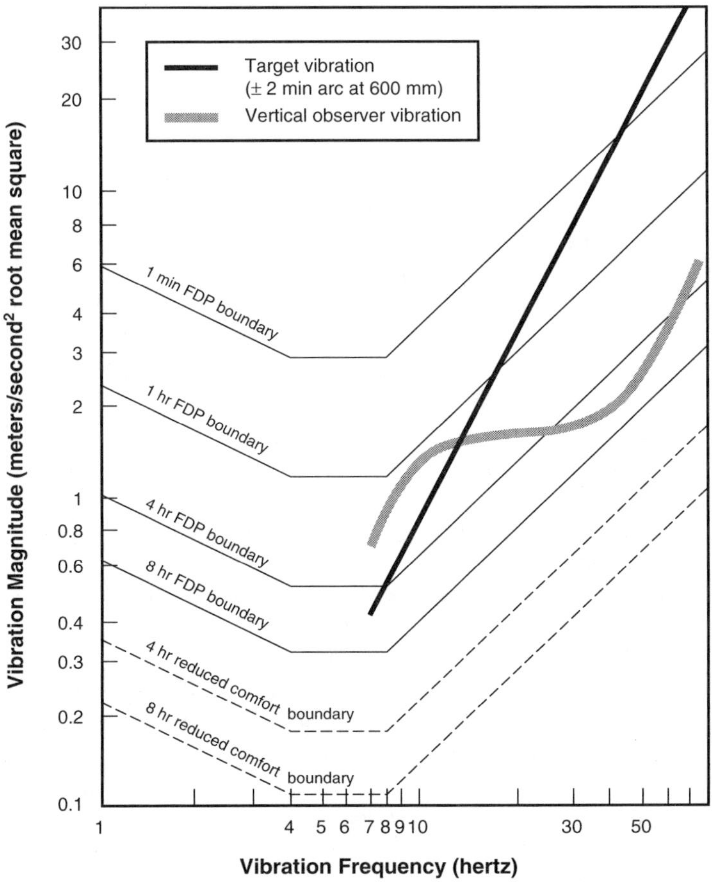

FIGURE 6-4 Quantitative limits on vibration for visual capability. FDP = fatigue decreased proficiency.

using the Land Warrior System are needed to establish the tolerance and reaction of soldiers wearing such equipment under operational conditions.

The Problem of Interaction

In approaches to understanding stress effects, one great limitation is the failure to investigate and capture the effects of multiple interactive sources of stress. Most detailed investigations have examined the influence of a single stress on a single type of performance task (e.g., Hancock, 1986b). Reliance on the arousal explanation enabled the postulation of a common pathway for all forms of bodily stress, as mediated through the ascending reticular activating system. This assumption has proved unreliable. Multiple stress effects cannot be inferred from the singular action of each one. A number of alternatives to the unitary arousal theory taxonomies offered by Hockey and Hamilton (1983) and Sanders (1983) can serve as a basis for ordering results, but they do not represent a causal account of the phenomenon at hand.

The major stumbling block in predicting stress effects concerns the interactive nature of stresses in the real world. The model by Hancock and Warm (1989) sees attention as the final arbitrator of performance efficiency and seeks to specify how multiple sources of stress influence this capacity. They distinguish two basic components of any source of stress: its spatial and temporal components. The temporal component involves the rate at which stimulation occurs, the spatial component involves the structure of the environmental source with respect to the perceiver. Since each form of stress must have these intrinsic characteristics, there is an opportunity to express different sources of stress on the same scale. This decomposition is easier for task stressors than it is for environmental stressors. However, since environmental stressors act through sensory systems in the same fundamental way as other stimulation to the central nervous system, the comparison is valid.

Extended research is clearly needed on specific systems in the operational context. Until the theoretical basis for predicting stress effects matures, there is simply no substitute for full-scale testing. It may be possible to forecast the effects of some specific forms of stress and to use these predictions as a basis for individual design. That strategy is probably more relevant to traditional ergonomic forms of stress such as heat, noise, vibration, etc., and their effects on performance and comfort than to more complex issues of software presentation and task design. Independent of original specifications, it is absolutely vital that user acceptance is evaluated at all stages of development of the Land Warrior System.

EFFECTS OF STRESS ON PERFORMANCE

Fatigue

Fatigue, a subset of general stress effects, is a concatenation of a variety of physiological and psychological factors. Among the precursors are extensive hours of work; among the physiological factors are circadian rhythms; and among the behavioral factors is vigilance or sustained attention. An individual experiences fatigue when the interaction of these multiple factors propels him or her across a "fatigue barrier."

Fatigue is a critical factor in military performance, especially for prolonged operations. Large and consistent efforts have been directed by the military and others to understand fatigue, yet it has resisted their best efforts. Attempts to define fatigue—for example, "subjectively experienced disinclination to continue performing the task at hand"—have to fight hard against the absurdity of tautology. In a classic work, Muscio (1921) even suggested abandoning the term *fatigue*, in part because of definition failure and in part because of its multidimensional nature.

Fatigue clearly has an impact on cognitive functioning. However, in the context of the infantry soldier, physical fatigue is equally significant. Perhaps the most important part of physical fatigue in terms of the helmet-mounted display is visual fatigue. Like any other component of the body, the eye is moved by muscles and these muscles are vulnerable to fatigue. Repetition of use usually brings on fatigue, although there is much evidence that even this putative muscular fatigue is actually mediated by central control. That is, a muscle that an individual judges to be fatigued will respond at almost 100 percent when stimulated externally. Because much of the information provided on the helmet-mounted display is expected to be visual, the question of visual fatigue during operations merits focused empirical effort. The physical fatigue the soldier experiences will not result solely from operating the equipment but is intrinsic to battle operations. The design of the display must therefore facilitate cognitive functioning under this form of stress.

Performance Failure

The human as an adaptive organism seeks to combat the adverse effects of stress for as long as possible. When such efforts are exhausted, the individual quickly begins to fail, both in performance and in physiological response. (A similar drop is evident in some aging individuals, who appear fit and active for many years, only to be taken extremely ill and die within a remarkably short time.) The question of relevance here is how to distinguish the onset of the drop. In complex task performance, the effects of stress are evident before they are measurable in terms of physiological response. It therefore makes sense to measure incipient individual failure *during* performance. Both subjective and physiological measures of cognitive workload (discussed in a later section) can be used to measure the onset of failure.

The speed and accuracy with which an individual accomplishes a task are the primary indicators of performance. In the process of failure, stimulus detection remains fairly rugged, whereas stimulus selection is more vulnerable to stress effects (Hancock, 1996b). One indication of incipient failure is that task-related cues are selected, but they are inappropriate for the action of the moment. For example, an infantry soldier may correctly detect and identify an enemy soldier at a distance, while a much closer enemy goes undetected. Evidence of a performance pattern such as this could signal to a platoon leader or other individual at a remoter site that performance is likely to degrade. Of course, such events need to occur in a pattern, since a single failure could occur for a variety of reasons. It is the job of the platoon leader to distinguish ongoing failure from momentary fluctuation in performance efficiency. This distinction is a difficult one to make and would depend directly on the current status of the engagement.

ADAPTIVE RESPONSE TO STRESS

The performance challenge for the soldier has traditionally been a physical one. Selection criteria reflect the capabilities needed to respond to physical demands, and training prepares the soldier to meet this challenge. A career in the military, ascending through the chain of command, has been traditionally accompanied by progressively greater responsibility and the burden of increasingly important decision making. Contemporary and proposed technologies are making changes in the challenge to the dismounted infantry soldier, in which significant cognitive demands are added to physical demands. The critical questions are whether soldiers will be able to meet this challenge and how to structure support systems that ensure success rather than impose failure.

Cognitive Workload

A vital step in the sequence is the assessment of cognitive demands. We note that the Army has experience in this area and has produced important information on workload assessment, in work led by Christ and his colleagues. However, it is important to recognize that the majority of information on cognitive workload has been derived from either laboratory experiments or field evaluations in which physical demand has played a nonsignificant or at most minor role. Much remains unknown about the effects on cognitive demand of physical activity and, conversely, the effects on physical activity of cognitive workload. Although comparable activities have been studied, such as firefighting and emergency mine egress, relatively little experimental information is available (but see Vercruyssen et al., 1988). Also, further research is needed on the interaction effects of trying to do several cognitive tasks at once. Although there is some evidence concerning dual-task performance and time-sharing capabilities, they have not been examined in association with physical effort. Much remains to be done in these areas.

The research literature includes almost limitless arguments on how to define cognitive workload.

For our purposes, we define workload[1] as:

time required/time available

---

[1]The major problem with this time-based approach to defining cognitive workload is the question of parallel processing. That is, individuals, especially skilled ones, can do more than two things at once. As a result, tasks overlap and it is difficult to provide a concrete estimate of time required, since tasks performed at the same time can appear to take "no" time.

If the time required by a task is more than the time available, there is cognitive overload. If the time required is very much less than the time available and the soldier has no other tasks to do, there is cognitive underload. Some tasks, such as extended surveillance, require the soldier to pay attention for long periods of time without overtly doing anything. This underload situation can be stressful and so result, almost paradoxically, in cognitive overload. Cognitive workload is best seen as a continuum on which both too much and too little are liable to result in problems.

The number and kinds of mental processes that can be carried out in a given amount of time are limited. For example, we can only listen to and comprehend one conversation at a time. Even well-practiced and automatic tasks, such as driving, can become difficult and attention-demanding. Encountering heavy traffic in an unfamiliar city may interrupt an ongoing conversation with a passenger or cause us to abandon various internal mental processes, such as imagery and planning. For the designer, understanding and measuring the attentional demands of various activities are important prerequisites for minimizing attentional overload.

Models of Attention

Broadbent (1958) presented one of the first attempts to characterize capacity limits in human information processing. He proposed that comprehending multiple, unrelated spoken passages was impossible because of limited short-term memory capacity. Therefore, some messages had to be selected for admission to short-term memory, while irrelevant information was essentially blocked by a filter mechanism. The filter could be set to admit information on the basis of physical characteristics such as location, pitch, loudness, etc. As Gopher and Donchin (1986) point out, Broadbent's model of attention does have implications for systems design. If it is important for an operator to focus attention on one source of information, it should be clearly segregated from other sources by simple physical features, to allow efficient selection by the filter. This same prescription applies to information conveyed by visual displays. Simple physical features, such as color, line orientation, and motion, can be picked up by preattentive processes (Treisman and Gelade, 1980; Wolfe, 1994) and can rapidly guide attention to relevant display areas. The filter model, however, has relatively little to say about how and when divided attention is possible. Nor does it account for the idea that tasks vary in difficulty, which can at least partially be overcome by increased effort.

Capacity Models

Kahneman's (1973) capacity theory of attention suggests that people have a limited supply of mental resources that can be divided among simultaneous mental activities. Difficult tasks require larger shares of resources and so cannot be carried out together without exceeding the supply. Easy tasks, well-practiced automatic ones, can be effectively "time-shared." The total pool of capacity is characterized as variable, depending on the subject's state as well as the task demands. When faced with a difficult task, arousal level may increase, providing additional resources to meet the increased demand. The increase in capacity can be monitored by measuring pupil diameter, which is related to arousal.

This framework led to a series of experiments by Beatty and colleagues (reviewed in Beatty, 1982). Their effort theory suggests that one can objectively measure task difficulty using arousal-related measures such as pupil diameter (see also Just and Carpenter, 1993). This model also points the way to new methods of measuring task difficulty based on the idea of spare capacity. A given task receives the resources necessary for performance; spare capacity can be measured by giving subjects a secondary task, providing an index of the capacity required to perform the primary task. In this way, the capacity demands of a variety of tasks can be measured and used to predict dual-task performance. Unfortunately, this endeavor did not meet with much success. Interference was often observed with even simple tasks, and the amount of interference seemed to be related to task similarity. For example, two visual tasks would interfere more than a visual and auditory task even though the "capacity demands" of the tasks were comparable. Kahneman had anticipated something like this, pointing out that "structural interference" between simultaneous tasks could occur when they both required a "non-shareable" mechanism. An example would be two tasks that simultaneously require eye movements to the left and right sides of a display screen. According to Kahneman, this interference is clearly different than that which arises from competition for a limited attentional capacity. As mentioned above, interaction effects can be overwhelming.

Multiple Resource Models

Wickens (1980) tried to account for the effects of task similarity on dual-task performance in terms of a multiple resource model. He suggested that resources correspond to a combination of three dichotomous dimensions consisting of: (1) stages of processing (perceptual/central versus response), (2) modalities of perception (visual versus auditory), and (3) codes used to represent and respond to information (verbal/vocal versus spatial/manual). Efficient time-sharing performance should occur when two tasks use different values on each of the three dimensions, as depicted in Figure 6-5. For example, retaining visually presented words calls on the visual input modality, central stages of processing (memory rehearsal), and verbal coding. Moving a joystick to track the location of a sound involves the auditory input modality, response level stages, and spatial codes. These two tasks should pose little interference when performed together.

Although debate continues on the number and nature of mental resources,

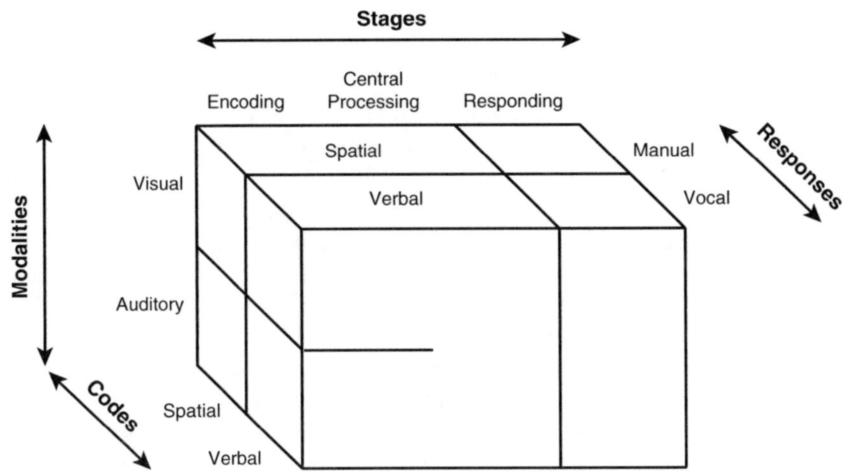

FIGURE 6-5 Proposed structure of processing resources. Source: Wickens (1984).

and indeed whether resources are a useful concept at all (Navon, 1984; Allport, 1993), the multiple resource model still offers a useful perspective on cognitive workload. First, it is clear that, if resources are of more than one kind, then workload assessments must measure each individually rather than offering a single, overall index. Second, the model offers a rough heuristic for designing task environments to minimize interference.

Recent research by Pashler (1989, 1994) has uncovered a surprisingly robust pattern of dual-task interference that occurs between even simple tasks. Suppose that subjects are required to make a speeded response to a tone by pressing a button. They are also required to make a rapid foot-press response to a visual signal that occurs shortly after the tone. When the interval between the tone and the light is small, the second response is delayed, in some cases by several hundred msec. This period of time, during which a second task is interfered with by a prior task, is known as the psychological refractory period. The magnitude of interference gets smaller and eventually disappears as the interval between the tasks increases. There appears to be a bottleneck in processing such that two tasks occurring in close temporal proximity are competing for access to a limited-capacity mechanism; the task arriving second must wait until this mechanism is no longer needed by the first task.

Pashler proposed that the particular pattern of interference he observed indicated that both tasks needed access to a single-channel mechanism, that is, a mechanism that can process only one input at a time. He further suggested that this single-channel bottleneck occurred fairly late in processing, at the point of selecting responses. The selection of the appropriate response for the second task had to wait until the response selection for the first task was completed. Similar effects have now been demonstrated for a wide variety of task pairings and response modalities, including hand and foot movements, vocalization, and eye movements (Pashler, 1994; Pashler et al., 1993). Carrier and Pashler (1995) have recently shown that retrieval of information from long-term memory, rather than response selection, is more generally responsible for the observed effects. Similar logic was used by Ruthruff et al. (1995) to show that mental rotation processes, which are used, for example, to imagine what a shape might look like in a different orientation, also use a bottleneck mechanism.

Thus there appears to be a mechanism involved with some aspects of memory retrieval and possibly memory transformation (as in mental rotation), which is strictly serial and constitutes a significant bottleneck in dual-task performance. It may be the same mechanism implicated in other theories that postulate a working-memory bottleneck. In addition, Pashler (1991) has shown that there is a second limited capacity attention system involved in perceptual processing. This system, unlike the central limited-capacity system, may be shareable by several perceptual inputs at a time.

There are obvious similarities between Pashler's model of dual-task interference and Wickens' multiple resource model. Both postulate separate limited capacity systems responsible for perceptual analysis and more central memory processes. Both systems allow for interference at stages involved with responding. According to Pashler, however, response selection presents a bottleneck regardless of the modality of the response system; verbal responses should interfere with verbal responses as much as with manual responses (Pashler, 1994). Pashler's proposals also bear some similarities to recent theories that point to working memory as the principal bottleneck in human information processing.

Working Memory Models

Knowledge in long-term memory is thought to be in an inactive state relative to the small subset of knowledge that is the focus of thought at any given moment. This activated knowledge is said to constitute working memory. A variety of cognitive activities, such as language, reasoning, and planning, require some knowledge to be maintained in an active state to be examined, transformed, and used for retrieval of additional information. According to a model proposed by Just and Carpenter (1992), maintenance of information in working memory requires attentional activation, as do other activities such as transformation and storage of working memory contents. This activation is drawn from a finite pool (similar to Kahneman's theory) and therefore, when comprehension becomes difficult, performance may slow down and become more error prone.

Just and Carpenter (1992) also propose that there are individual differences in working memory capacity as measured by the working memory span test, that have implications for how well an individual comprehends a difficult instruction. For example, they found that low-span and high-span subjects read simple sentences at about the same rate, but that low-span subjects had particular difficulty reading and comprehending syntactically difficult sentences. Similarly, St. George et al. (1995) found that high-span subjects were more likely to make optional inferences in reading a text.

Interestingly, pupil diameter has been shown to be sensitive to working memory load in a variety of tasks, including reading (Just and Carpenter, 1993) and problem solving (Just et al., in press). This is consistent with Kahneman's original claim that pupil diameter may serve as a general measure of processing difficulty. It remains to be seen whether working memory will prove to be fractionated into separate limited-capacity systems, as occurred with Kahneman's model. For example, Shah and Miyake (in press) have recently provided evidence that there may be separate working memories for language and visual/spatial tasks, which is consistent with recent brain imaging work showing that these tasks activate separate brain areas.

Measures of Cognitive Workload

As the previous sections suggest, there are severe limits in our ability to process multiple sources of information. Understanding these limits is essential in designing and evaluating components of the Land Warrior System. Current technology allows many options in terms of information display, graphical interface, input devices, etc. The availability of these options has the potential to produce severe information overload for an infantry soldier who is hot or cold, tired, and stressed. Comprehensive assessment of workload is therefore a critical part of the design process.

The goal of cognitive workload measurement is to characterize the attentional demands that a task places on an operator. O'Donnell and Eggemeier (1986) suggest several criteria for evaluating measures of cognitive workload. Measures can be evaluated according to their: sensitivity (does the measure respond to variations in task difficulty or load?), diagnosticity (does it indicate what kind of attentional resource is being used?), intrusiveness (does the instrument interfere with performance of the task?), implementation requirements (does it require costly equipment or large amounts of time to complete?), and operator acceptance.

Table 6-1 lists four classes of workload measurement with their pros and cons. Primary task performance and subjective measures can be used for the initial screening of designs. These are easy to administer, sensitive to task difficulty, and have good user acceptance. Subjective measures include structured interviews, rating scales focused on particular tasks, etc. They can be supplemented by looking at how easy it is for operators to combine various tasks together, using the embedded secondary task technique. Dual-task performance, using pairs of tasks that are likely to occur together in the course of a mission, should provide valuable information on potential bottlenecks.

TABLE 6-1 Workload Measurement Methods

Method	Pros	Cons
Primary task	High face validity Measures assessed anyway Nonintrusive	Workload or performance? No productivity
Secondary task	Good diagnosticity Sensitive Poor user acceptance	Intrusive Loading task in high situations Theory bound interpretation
Subjective response	High face validity User acceptable Easy to obtain Interscale replicability	Dissociation with primary Largely post-hoc measures
Physiological assessment	Mostly nonintrusive Objective Data rich Globally diagnostic	Subject to artifacts Poor user acceptance Difficult to administer

Several useful reviews of these techniques are available. Gopher and Donchin (1986) provide an overview of the concept of workload, O'Donnell and Eggemeier (1986) give a good description of various subjective scales of workload measurement, and Kramer (1991) has provided a recent analysis of physiological workload measures. In the section that follow we summarize the important points and consider their relevance to the Land Warrior System.

Subjective Measures

There is a subjective aspect to attention. We are often aware of increases in the difficulty of tasks as well as our own efforts to compensate for that difficulty. This effort or intensity dimension of attention is the basis of Kahneman's model and its many successors. Subjective measures are an attempt to systematically query subjects about their own awareness of task difficulty. The Cooper-Harper scale is one of the earliest efforts to assess workload. Pilots are given a questionnaire that requires them to rate the handling characteristics of aircraft, as well as how much compensation is required to make up for handling deficiencies. Later versions were modified to specifically measure workload and were found to be highly correlated with a variety of task difficulty manipulations (North et al., 1979). Clearly, this scale attempts to provide a general or global measure of workload without being diagnostic as what factors are driving the workload.

Task difficulty is perhaps the most obvious contributor to subjective workload, but it is certainly not the only one. A task that appears easy in one case can be made difficult by applying time pressure. So part of workload involves a sense of time pressure. In addition, physical demands may contribute to our sense of mental workload. Two recently developed instruments provide subscales to measure some of these different aspects of workload. The NASA task load index or TLX (Hart and Staveland, 1988) asks operators to make ratings on six dimensions: mental demands, physical demands, temporal demands, performance, effort, and frustration. A weighted average of scale values on these six dimensions is computed to provide an overall measure of workload. Like the Cooper-Harper scale, scores on the TLX show high correlations with various criterion variables, such as performance in simulators as well as in field tests (Hart, 1986). The subjective workload assessment technique (SWAT) uses three scales designed to measure time load, mental effort load, and psychological stress load. For example, operators are asked to estimate mental effort in terms of whether the task required little conscious mental effort, moderate effort, or extensive mental effort and concentration (O'Donnell and Eggemeier, 1986). Weights are then assigned to the three scale values to derive a composite measure of workload.

The multidimensional nature of SWAT and TLX appear to offer an increase in diagnosticity relative to single-scale approaches such as the Cooper-Harper scale. For example, one can look at the scale weights for a given subject to determine whether that subject perceived the task to be time stressed or high in frustration. On the negative side, these scales may take a long time to administer, and, in a complex system like the Land Warrior, with many components, this time could be prohibitive. In addition, the meaning of the weights assigned to the dimensions is not perfectly clear. In comparing TLX and SWAT, Nygren (1991) points out that equal weighting of the scales in TLX ought to be about as good a predictor of the criterion variable as the differential weights derived from the data, and in fact that has been found empirically (Byers et al., 1989).

It also appears that subjective measures may sometimes dissociate from other measures of workload. For example, Sirevaag et al. (1993) collected TLX ratings from helicopter pilots in a high-fidelity simulator. They varied communication demands and found that pilots had trouble adhering to nap-of-the-earth altitude criteria under high communication demands. The greater load imposed by the communication task was reflected in several workload measures but not in the subjective ratings. When questioned about this puzzling outcome, pilots indicated that they were aware of the greater difficulty involved in the demanding communication condition, but their ratings didn't reflect this because they felt that none of the conditions exceeded their capacity to perform successfully, and so they rated them as equivalent. Subjective measures are just that, and subjects can adopt their own criteria for the ratings.

Hendy et al. (1993) point out that both the TLX and SWAT go to great lengths to provide a composite measure of workload. They suggest that, if one is interested in a global measure of workload, one might do just as well by asking subjects to simply estimate the workload on a univariate scale. The procedure used by SWAT and TLX seems to assume that, although subjects may be able to give accurate estimates on specific components, they are not able to report on overall or global workload. In four studies, Hendy et al. compared the composite score on a modified TLX test with a univariate measure of global workload, asking subjects to use a magnitude estimation procedure to estimate the difficulty of various segments of a flight relative to the difficulty of the takeoff and departure segments. They found that the univariate measure was more sensitive to variations in task difficulty than the TLX composite measure or any of its subscale scores, suggesting that, if a global measure of workload is what is desired, a simple univariate scale works pretty well. They also raise an important issue about the goal of workload measurement: How useful is it for the system designer to know that a workload problem stems from excessive mental effort, time pressure, or frustration? They also point out that "excessive time pressure could come from the number of things to attend to, the requirement for high precision, lack of feedback, or use of inefficient strategies attributable to insufficient training" (p. 599). In other words, it might be difficult for the system designer to use the kind of knowledge provided by workload measures to diagnose what aspect of the design should be changed.

Hendy et al. (1993) recommend presenting scales that focus on various aspects of the design in terms of impact on attentional demands. For example, knowing that memory retrieval of unfamiliar information is taxing or that visual clutter requires perceptual resources, one can examine various aspects of the Land Warrior System that appear to have these characteristics and evaluate different designs in terms of perceived workload. Subjective methods that are attuned to aspects of the particular design (e.g., evaluating the cursor control, using the helmet-mounted display to view maps in a navigation exercise) and are motivated by a theoretical understanding of attentional systems (such as the multiple resource model) may provide a good first-pass estimate of which aspects of the design present likely workload bottlenecks.

Primary and Secondary Task Measures

Primary task measures simply try to infer workload from performance on the task of interest. Primary task performance is obviously the critical variable. However, it isn't clear that primary task measures have much of a direct association with workload. Errors in performance do not necessarily indicate high workload imposed by the primary task. They can arise from a variety of sources, including workload levels that are too low, as in vigilance tasks in which the operator may miss signals because they are so infrequent.

A better measure of workload is provided by secondary task performance, in which spare capacity is assessed by presenting operators with occasional probe signals that require them to press a key. Probe response time should be related to the difficulty of the primary task. For example, the difficulty of choosing various options of map presentation on the helmet-mounted display could be evaluated in terms of speed of response to occasional auditory tones.

We pointed out earlier that there are a number of difficulties associated with using secondary task performance as a measure of spare capacity. Subjects may use various strategies for dealing with what is basically a dual-task situation; some may actively prepare for the probe task despite instructions to treat it as secondary. In addition, the existence of multiple resources means that a given probe task, such as auditory detection, will vary in difficulty depending on its similarity to the primary task (due to variation in overlap of the particular resources used by each task). Finally, the secondary task can be intrusive and therefore disruptive of primary task performance. As we noted in connection with the model of dual-task interference (Pashler, 1994), even very simple and dissimilar tasks can produce interference when both of them require memory retrieval at the same time. These considerations suggest that secondary task performance should be viewed with caution as an index of workload.

Of course, soldiers using Land Warrior equipment will often be in dual-task situations. For example, a soldier may be navigating terrain with the aid of the map display and GPS when an auditory message comes in. The message has to be checked for its importance relative to the navigation task; the speed and accuracy of response to such messages would be therefore expected to be a function of the ease of use of the map system. This is an example of an embedded secondary task method, in which the tasks are presented within the context of meaningful scenarios that are motivated by the kinds of dual-task combinations that are likely to occur in operational use. The panel recommends that this sort of dual-task analysis be carried out using a variety of realistic task combinations. At the very least, this approach would provide valuable information on which kinds of dual-task situations pose difficulty. It could prove valuable in design as well as test and evaluation.

Physiological Measures

Measures of brain activity such as the electroencephalogram (EEG), have potential to reflect cognitive workload. Averaging procedures can be used to derive the event-related potential (ERP), which reflects electrical activity associated with a particular signal. For example, subjects can be required to count "target" tones of a particular pitch embedded in a series of nontarget tones having a different pitch. The targets will be associated with a particular component of the ERP known as the P300. The amplitude of the P300 is associated with how much attention the subject allocates to the signal (Israel et al., 1980). As the primary task becomes more difficult and requires more attention, the P300 associated with the target tones is reduced. This kind of trade-off between tasks in terms of P300 amplitude has been demonstrated in several studies (e.g., Hoffman et al., 1985; Israel et al., 1980).

Measuring the ERPs of probes is clearly a variation on the secondary task method, and it is fair to ask whether the information provided by this technique is worth the additional expense and complexity associated with EEG recording. One advantage of the P300 relative to reaction time measures is that it appears to be sensitive to perceptual/central processes and is not affected by the response system. Reaction time, in contrast, is sensitive to limited capacity processes from input to response. P300 therefore offers an increase in diagnosticity over behavioral measures.

As we have pointed out, the secondary task method can be modified by using an embedded task. This is also true of the P300. A good example is provided by Humphrey and Kramer (1994), who presented a dual-task situation in which operators monitored a series of gauges for critical readings. In addition, arithmetic problems could be periodically presented on the same screen used to present the gauges. The difficulty of each task was manipulated and P300s collected in response to the presentation of information in each task. The goal was to determine whether it was possible to determine the difficulty level (and presumably workload) that subjects were experiencing by examining P300 amplitude. They found that 90 percent discrimination accuracy could be obtained by using 1 to 11 sec of ERP data. These results suggest that the P300 has the potential to serve as a real time, diagnostic index of momentary fluctuations in workload.

An additional potential application of EEG methodology is relevant to the Land Warrior System. Work at the Naval Health Research Center by Makeig and Jung (in press) has shown that EEG can be used to monitor alertness. They found that they could accurately predict, in real time, the likelihood that operators would miss occasional sonar signals by monitoring EEG power in particular frequency bands. When operators enter states of low alertness, they could be warned by a monitoring system. Given advances in the development of dry electrodes and miniature amplifiers, this development raises the possibility that real-time alertness monitoring could be included in the Land Warrior System for soldiers in vigilance situations.

There are other physiological measures available as well. For example, the electro-oculogram provides a measure of eyeblink frequency and duration over a given period of time. Blink rate frequency and blink duration both tend to decrease with higher levels of workload (Stern and Skelly, 1984). Electrocardiogram measurement quantifies a subject's heart rate and heart rate variability. Although the average heart rate seems to be generally insensitive to small or moderate fluctuations in workload (it may rise with stress, however), the variability of the heart beat (sinus arrhythmia) can be a good indication of mental workload. In general, the heart functions with less variability between beats at higher levels of workload. Various filtering techniques have been developed to increase the sensitivity of this measure by reducing noise due to other bodily functions (Mulder, 1980; Porges, 1985; Moray et al., 1986).

Model-Based Approaches

In this section, we review recent attempts to develop detailed models of task performance that can provide insights into limitations of dual-task performance that would be difficult to obtain using the measures reviewed above. A thorough understanding of task interference and competition for limited capacity mechanisms will probably require a detailed simulation of how humans perform a given task. Such a simulation should provide a detailed account of the order, duration, and capacity limits of the elementary cognitive processes assembled for a given task. The model human processor (MHP) proposed by Card et al. (1983) is a good example of this approach. They specified a set of elementary information-processing mechanisms (for example, making an eye movement), along with associated execution times. A detailed task analysis then made it possible to specify the particular arrangement of subprocesses involved and arrive at an estimate of the total time required to perform a given task. The MHP does a fairly good job of predicting these times, but mainly for simple tasks (Eberts, 1994). With increasing complexity, the job of specifying the path of elementary processes becomes quite difficult.

A modern predictive modeling approach, similar in spirit to the MHP, is the EPIC model (executive process-interactive control) of Kieras and Meyer (1995). Their simulation system specifies a number of input systems (eye, ear, etc.) and output systems (manual, vocal, eye movement, etc.). In general, EPIC contains a much richer set of input and output devices than MHP and can therefore be applied to a wider range of problems. In addition, decisions, transformations, and such are carried out by a set of production rules operating in working memory. A simulation is performed by first specifying the algorithm by which the task is accomplished in terms of the set of production rules required. Exposure to the task domain initiates a cycle of "firing" of the production rules and changes in input and output devices that generates the desired behavior. We note that, although the experimenter has to specify the algorithm by which the task is accomplished, this ought to be easier than specifying the detailed chain of processes required by MHP.

EPIC has been applied to the simulated task of a telephone operator interacting with a workstation to help a customer complete a call. The model did a good job of predicting the total time on task as well as times for individual keystrokes by the operator. In addition, the model showed that the major limitation on operator speed was not typing time but the rate at which the customer spoke the telephone number. These instances of model-based insights into the nature of bottlenecks illustrate that this approach can be quite useful in redesigning tasks and interfaces in an attempt to minimize interference. Creating detailed models is a time-consuming process and requires a thorough task analysis. This probably would not be feasible for every soldier task, but it could be worthwhile for particular subsets of tasks that are critical to a mission. In the case of the telephone company, saving a few seconds on each call can result in savings measured in the millions. Savings of this magnitude on the battlefield could be the difference between mission success and failure.

Unresolved Problems

As noted throughout this chapter, there are many unresolved problems concerning the assessment and interpretation of cognitive workload and its demands. Already mentioned are the unknown effects of combining physical and mental demands. In addition, many other questions have to be faced in using this form of assessment. For example, there is very little information on the differences among individuals (Damos, 1988). Furthermore, the effects of training on mental workload are still to be clarified, especially when training takes place under conditions that only simulate actual operational environments (see Hancock et al., 1994). The effects of task failure on workload are uncertain (Hancock, 1989), and strategies to improve performance through adaptive or compensatory systems have only recently begun to provide potential answers (Chignell and Hancock, 1985; Hancock and Chignell, 1987; Hancock et al., 1994). Also unknown are the relationships among stress, workload, and other aspects of performance, such as situation awareness. In particular, there is concern that, under certain conditions, workload and performance dissociate, that is workload increases but performance gets better or, similarly, workload decreases but performance worsens (Derrick, 1988; Yeh and Wickens, 1988). This is especially disturbing for those who want to use cognitive assessment of workload as a basis for design decisions and the definition of operational procedures. As technical support systems are being developed, these become vital experimental questions to be addressed.

What makes the job of the soldier different from, say, a pilot are the extreme physical demands placed on the individual. Indeed, the infantry soldier has to carry his technology with him and, unlike others, is not carried by it. The soldier has to meet the demands imposed by operational conditions, be it the climate or the theater of operations or simply the need to maintain mobility over difficult terrain.

We cannot overemphasize the significance of the fact that the lessons learned so far about human interaction with complex technologies have been garnered under quite sedate conditions. Typically, the individual is well-rested and seated in a light, well-ventilated, and comfortable situation and then asked to perform the necessary tasks. We know little about perception, action, and complex decision making during or immediately following strong physical exertion. It is essential that such testing be initiated prior to system design. One recommendation that can be made now is for tasks to be presented, not in the typical alphanumeric form but rather in graphic form. The latter are more easily and reliably solved and are better able to resist the deleterious effects of stress (Hancock, 1996b).

As we have noted in the discussion of stress, one critical characteristic is the reduction of the range of cue utilization (Easterbrook, 1959). This phenomenon has become to be known as *narrowing*. Although difficult to show in the laboratory, in the real world, it is a common experience. Most people can recount an incident, such as driving in bad weather, when their whole attention was focused on a single object, such as the road ahead. In experimental work, Hancock and Dirkin (1983) have shown that this is an attentional strategy, rather than a visual phenomenon (such as occurs in the restricted vision of pilots under large G-loads). As a consequence, extreme stress can restrict attentional scanning of other environmental sources.

If the source of information to which the soldier narrows is located on the helmet-mounted display, he may become "blind" to other threats in the environment. Consequently, what is displayed to the soldier during action is a critical decision; it may be advisable to provide some cue to change attention from the information displayed in the helmet to that available externally. Ideally, these two sources would be fused so that conflicts do not occur. It is also clear that narrowing relates in some degree to the notion of situation awareness. A strong research effort is needed on the phenomenon of narrowing, especially as it relates to soldier activity involving helmet-mounted displays for advanced tactical engagements.

Improving Adaptive Response

In general, the adverse effects of stress are very difficult to combat. Repeated practice and training in appropriate conditions do provide some protection from the worst aspects of stress. However, they are far from achieving a situation in which stress does not influence performance—or even improves performance skills.

Accounts of battles in the eighteenth and early nineteenth centuries emphasize the confusion of the battlefield and the reliance on hearing rather than vision, which was totally obscured by the smoke of artillery. Confusion is an ally to the enemy. Confusion has often lost battles, and certainly it is a source of extreme stress to the infantry soldier. Therefore, any technology that serves to reduce confusion is at one and the same time helping mission performance and reducing the incidence of stress.

One factor that has been observed to affect the response to stress is the perceived degree of control over the situation. Besides reducing confusion, control dissipates stress. An individual who feels in control of a situation, is less likely to experience stress in that situation. It is anticipated that new helmet-mounted display technologies will facilitate the distribution of control from a single operational headquarters to individual squads and even members of those squads. On the modern battlefield, decisions are directed to those with the most relevant information, on occasion, that will be the front-line soldier. This added sense of control will be a critical influence in reducing stress.

Information About Battlefield Context

There is a tendency in many technological systems today toward an ever-greater degree of complexity. By complexity, we specifically mean the number of degrees of freedom in the system, which translates to the number of potential states of the system that can be communicated to the operator. With pull-down menus, window systems, and direct database accessing, it will be technically feasible in the near future to give the infantry soldier almost unlimited on-line access to all aspects of human knowledge.

Such a prospect is the paramount example of a plethora of data but no information. Information in this context is data that are directly applicable to the immediate conditions in which the soldier finds himself. On a battlefield of uncertainty, the aim of such support technology should be to reduce uncertainty, not increase it. Thus the challenge to design is to provide the appropriate information in its simplest form, just as it is needed.

One potential solution to this question lies in the framing of mission objects as perceptual-motor demands, not esoteric cognitive operations. In a recent article, Hancock (1996a) has indicated how such a metaphoric representation could be achieved. This approach would essentially superimpose a picture of the demands on the real-world view. This would minimize alphanumeric presentation and emphasize data fusion between the real and task-represented worlds. By simplifying the data presentation format, the probability that the system will be successfully deployed and operated in the stress of battle would be significantly increased.

These observations relate to the actions of the individual soldier in the operational environment. However, given the structure of the forces as currently configured, the additive contribution of information from each and every soldier to the next higher level of command promises to increase workload geometrically at each succeeding level. Consequently, if information is not filtered in some fashion, made context-contingent in some fashion, and fused in some fashion, higher levels if command will be blinded not by the paucity of data but by an overwhelming overload of it. Consequently, as this program moves forward, it is important to consider workload issues beyond the individual soldier and to address specifically how the evolving architecture of the Land Warrior technology can manipulate information at even higher levels of command to achieve mission goals.

Training, Expertise, and Individual Differences

However intuitively logical the display may first appear, training will be needed to maximize the use of this new form of soldier support. In accordance with the observations of Schneider (1985), the training should emphasize the consistency of representation and the linkage between representation and real-world actions—for example, always designating the enemy by the color red. With such training, it is possible that low-level attributes of stimuli used in the display, can be processed in a speedy and error-free manner. Furthermore, the extensive use of such consistency would render task performance less vulnerable to the stress of uncertain conditions (see Hancock, 1986a).

Soldiers' facility with the use of helmet-mounted displays and other advanced technologies can be expected to vary greatly among individuals. For example, some recruits from rural areas may not have much experience with comparable technologies, such as video games. Others will have spent many hours on technologies that, although designed as games, may have a strong transfer to the new operational equipment. Consequently, software configurations for initial training need to be designed with this baseline performance difference in mind.

Although individuals can be selected for specific operations, what remains to be clarified are the intrinsic abilities (e.g., superior spatial orientation) that would make some soldiers immediate experts at the task presented via a helmet-mounted display. There is a body of knowledge concerning soldier skills and initial screening batteries for performance capability; however, the proposed changes in soldier function are so great as to require further research evaluations of these capability screening issues. An initial software consideration is how to structure the displays so that soldiers find a challenge in their repeated performance and thus, like video game players, become highly proficient at their task. This requires some knowledge of task motivation and especially how to train for performance skills under extremes of stress. In essence, the problem of individual differences in response to cognitive workload—one of the most underresearched areas in all human factors—is critical to performance success. More knowledge is needed in this area to ensure the success of the new technologies.

IMPLICATIONS FOR DESIGN

We can never lose sight of the infantry soldier's primary goal of mission accomplishment. In designing a technical support system to be most conducive to effective performance, at no point should such a system hinder this primary objective. The infantry soldier must recognize the helmet-mounted display as a vital piece of his equipment that is critical to survival. Equipment that is cumbersome, unreliable, or ineffective can and will be discarded in the extremes of battle. Consequently, the thought at the forefront of design should not be the feasibility of implementation but everyday utility to the individual soldier. If the helmet-mounted display does not work well in the stress of battle, burdening the soldier with advanced technology systems is a clear disservice.

In our consideration of the Land Warrior System's helmet-mounted display, which will be used to display maps, symbology, sensor images, and other information, we have reviewed recent work in psychology directed toward reducing the cognitive workload associated with visual displays of information. Based on visual and cognitive limitations that can make information hard to see and interpret, Wickens (1992) offers a useful list of principles of display design, along with examples. We duplicate that list here and examine a few of these principles in more detail.

Perceptual Principles

1. Absolute judgment. Don't require observers to make judgments on a variable such as color or brightness using more than 5-7 levels.

2. Top-down processing. Don't violate expectations based on past experience.

3. Redundancy gain. Presenting the same message more than once, particularly in different formats (e.g., voice and visual display) increases comprehension.

4. Similarity. Similarity can cause confusion, both in perception and in memory. A useful measure of similarity is the ratio of similar to dissimilar features.

Mental Model Principles

5. Pictorial realism. A display should look like the variable it represents.

6. Moving parts. Dynamic aspects of displays should move in accordance with the user's mental model of what is being represented. For example, an altimeter indicator should move up with increases in altitude.

7. Ecological interface design. Displays should bear a close correspondence to the environment being represented. Adhering to the pictorial realism and moving part principles should help one achieve pictorial realism.

8. Minimize information access cost. Disengaging attention and the eye from one display location and moving to another require time. Information should be located in such a way that access time for frequently used information is minimized. This principle has obvious applications in design of menus for computer interfaces.

Attentional Principles

9. Proximity compatibility. Sources of information that must be integrated

should be close together in the display. In some cases they can be integrated into a single "object." Conversely, sources that don't require integration should be spaced to enhance focused attention on each source.

10. Multiple resources. Dividing one's attention is easier when information uses separate resource pools, e.g., vision and audition.

11. Predictive aiding. When possible, have the system aid in predicting future values. Limits in working memory mean that operators will often be busy processing current information and will fail to project what will happen in the future.

12. Knowledge in the world. One way to reduce memory load is to place required information in the environment. A pilot's checklist is an example.

13. Consistency. As much as possible, displays should be consistent with recently viewed displays or habits that the observer brings to the situation.

As Wickens points out, some of these principles conflict with each other. The proximity compatibility principle is a good example. Wickens and Carswell (1995) distinguish between display proximity and processing proximity. Display proximity refers to the distance between two sources of information in the user's perceptual space. Proximity here refers to more than simple physical distance. Various grouping principles such as common color, motion, and shape can unite display elements that are separated spatially. Proximity can be enhanced by combining sources into a single "object." For example, two variables could be represented as the height and width of a rectangle. Their product could then be directly perceived as the area of the rectangle.

Processing proximity refers to whether or not the operator needs to combine information from different sources. An example of high task proximity would be a process-monitoring task requiring the operator to integrate two variables, such as rate and duration. In other cases, the operator might have to monitor two sources that are independent and therefore have low processing proximity. Compatibility between these two ways of defining proximity is clearly desirable. Close processing proximity can be enhanced by close display proximity and similarly for distant processing and display proximity. Wickens and Carswell describe several display techniques that are designed to manipulate display proximity and achieve the desired goals in terms of the operator's task.

The principle of display proximity is particularly relevant to the helmet-mounted display. Wickens and Carswell (1995) point out that head-up displays in aircraft usually lead to better performance than the older head-down format. One clear advantage is display proximity. The symbology presented on the head-up display (near domain) is closer to information in the outside world (far domain) than is the case with the head-down format, producing an advantage in terms of access costs. This advantage is even greater for near and far domain information that is "conformal." This follows from the definition of display proximity. Making information conformal helps integrate information across the two domains. In contrast, information in the two domains that is nonconformal will tend to be segregated, enhanced focused attention. This may be one reason that attention to head-up display symbology may sometimes result in missing rare events in the far domain.

Many of these same principles can be used to guide the creation of maps and graphs that are easy to comprehend. Detailed recommendations for these domains can be found in Kosslyn (1994). A recent book by Travis (1991) provides guidelines for the effective use of color in displays. Shah and Carpenter (1995) present several studies aimed at determining how subjects represent graphical information in working memory.

CONCLUSIONS

This review makes clear the insufficient state of knowledge about the stress and cognitive workload of individuals performing physical and cognitive tasks at the same time. Although there is extensive information about physical effort and physical workload, as well as some insight into cognitive workload, we do not yet know enough about their combined effects to make definitive pronouncements relevant to future infantry operations. One reason is that current models of stress and performance are insufficient to determine how combined physical and mental effort may affect performance. Clearly, a strong effort is needed in basic stress and performance research.

Another critical question in need of more research is how to provide the soldier with task-related information and how to suppress extraneous information so as to avoid information overload. Even with the best designed equipment, excessive information presented via a helmet-mounted display threatens to induce "cognitive capture," in which the individual becomes oblivious to the threats of the external environment. Information must be presented in a way that does not dominate the individual's attention. This clearly relates to research on situation awareness, as discussed in Chapter 3.

Finally, better understanding is needed of the individual differences in responses to cognitive workload and the relation of those responses to standardized training strategies. One strategy for dealing with such differences mentioned in this chapter is the use of customized adaptive interfaces that can be tuned to the particular user. However, we still need to know more of what drives individual perceptions of load, especially under operational conditions.

DESIGN GUIDELINES

1. Whenever possible, provide graphic representations of the problem, since graphics are processed more efficiently and accurately under stress.

2. Whenever possible, simplify the graphic representation of the task do-

main, since the task itself is a stress in addition to the threat conditions. Simplification of decisions will reduce workload in high-load situations.

3. Present salient information in the center of the visual display. Information on the periphery of the display will be lost in high-stress conditions.

4. Reserve the presentation of complex alphanumeric information for pre-mission phases and post-mission debriefings unless absolutely necessary.

5. Whenever possible, use redundant auditory and visual warnings for threat location.

6. Whenever possible, use visual presentations for detailed communication of information.

7. Provide global help functions (e.g., location of nearest friendly force) at all times.

8. Reduce menu and data entry requirements to a minimum, especially during engagements.

9. Provide a restricted, understandable icon set to reduce stress in processing.

7

Research, Development, Testing, and Evaluation

The Army has a well-defined developmental and operational testing program for all new major and minor systems. The complex Land Warrior System, with its range of evolving technologies embedded in the helmet-mounted display, requires even more advanced test programs, supported by the thorough identification of critical issues and criteria and driven by systematically developed measures of effectiveness and measures of performance. In order to achieve a smooth testing process for this complex program, specific performance criteria—which are not presently available—must be identified to allow for unambiguous subsystem evaluation prior to testing the complete, integrated Land Warrior System.

With regard to the design effort for the helmet-mounted display, the lack of user-identified performance priorities makes the development of an effective test and evaluation program more difficult. Key issues in building an effective research program and an effective test and evaluation plan include identification of the technical areas requiring additional research and performance requirements that need discrete, objective evaluation. Historically, human factors research in the Army's programs has been underfunded and understaffed; it has not been a primary driver of revisions to proposed requirements; and it has not played a central role in the planning of creative testing programs. This situation needs to be changed.

In this chapter, we highlight the research that the panel judges to be essential and present a concept for research and testing that effectively incorporates critical human factors considerations for the Land Warrior System's helmet-mounted display.

In order to properly address these issues, a three-tier, highly integrated research, testing, and evaluation strategy is proposed. This approach represents a shift from current practice in that it includes an intermediate, semi-controlled set of research and testing experiments between laboratory and bench testing and operational field operations. Equally important, it incorporates the active involvement of users in every stage in the development sequence.

First, do the technologies and capabilities embedded in the helmet-mounted display provide a significant improvement in the soldier's local situation awareness under the expected environmental and operational conditions? Second, does the helmet-mounted display significantly improve the visual perception of a soldier and his ability to perform operational tasks under tactical operational conditions? Third, how does the cognitive workload created by the helmet-mounted display affect task performance by soldiers under tactical and operational conditions? And forth, can current infantry soldiers, with existing skill qualifications, perform their expected operational tasks more effectively with the capabilities provided by the helmet-mounted display? These issues are elaborated in a later section of this chapter.

ADOPTING A NEW STRATEGY

Until the advent of the electronic age, the systems a soldier was required to use were relatively simple mechanical devices, and field testing had only to demonstrate that the equipment would continue to function in mud, rain, and ice. As systems have become more complex, requiring sophisticated training, and electronics-intensive with information transfer aspects, the gap between laboratory or bench testing and operational testing has become much larger. As a result, it is now possible for a system to meet all its engineering specifications in laboratory tests but turn out to be unusable in the field. Furthermore, the reasons for failure in the field may be obscure to the engineering design team, leading to "chase the tumbleweed" redesign efforts. The problem is most likely to be related to the combination of variations in field conditions and variations in the attributes of human users. The failure to recognize and deal with these variables and their interaction can result in incorrect definition or incomplete specifications that lead to limited prospects for success.

We propose a methodology that brings the user into the testing and evaluation process earlier, through controlled testing that combines some varied environmental and personnel conditions from operational testing with the structured data collection and controlled conditions characteristic of laboratory testing. It is proposed that testing and evaluation for the helmet-mounted display be developed in three tiers as follows:

1. Controlled laboratory or bench testing of technical performance of the system—both with and without human users.

2. Controlled field experiments with a variety of users, from experienced to new entry, and with system experts from the design team involved.

3. Operational test and evaluation exercises employing soldiers from the target population in virtual-type simulations and live simulations in a realistic operational environment.

The insertion of the mid-level tier of trials allows for the interaction of potential users and the design team in conditions that combine structured data collection with variability in environmental conditions (e.g., day, dusk, night for visual factors; camouflage for terrain variations) and individual variation in users (e.g., effects of regional accents on the performance of a voice recognition system for acoustics).

We list below some guidelines, considerations, reservations, and underlying problems in human performance measurement from the *Guide to Human Performance Measurement* (American National Standards Institute) that should be kept in mind in designing the testing and evaluation program and user trials:

- *Lack of a general theory to guide performance measurement.* At present, only a few basic relationships have been discovered—for example, the inverse relationship between speed of task performance and accuracy, or performance quality.
- *Measurement control and realism.* An inverse relationship exists between the control of measurement conditions and operational realism.
- *Behavior is multidimensional.* Many factors influence human performance, some of great weight, others of little importance. Moreover, these factors probably change their importance over time and with different measurement contexts, and they may have different weights in different individuals. This results in the need to perform more experiments with more variables, subjects, and test trials.
- *Unclear relationship between objective and subjective data.* Objective measures are based on observable behavior (i.e., moving a lever, speaking a word). For complex tasks as well as phenomena that are not directly observable, objective measurements may not be possible or may be too costly to develop.
- *Generalizability.* Results may not be generalizable to the real world.
- *Measuring cognitive activity.* Cognitive activity is inherently more difficult to measure than physical performance because it takes place within the individual. Cognitive activity cannot be observed directly; it requires analysis of the output consequences of the cognition and, even more important, some kind of self-reporting, which may be tainted by its subjectivity. This means that measurement of behavior continues to be an inherently difficult process that is not becoming easier. In these cases, performance must be inferred from system inputs, outputs, and states, as well as from models of the operator's cognitive processes (Vreuls and Obermayer, 1985).

• *Performance criteria.* There is a lack of objective performance criteria for most tasks. This lack of criteria makes it difficult to (a) assess performance quality and sufficiency and (b) identify the type of performance measurement techniques required to operationally define significant differences in performance (Vreuls and Obermayer, 1985).

The considerations presented above are not intended to reflect an inability to design and evaluate user trials but rather to point out some potential pitfalls. Later in the chapter, we list the test conditions and criteria that may be useful in any testing or evaluation sequence designed to assess the helmet-mounted display and the Land Warrior System.

PLANNING THE STRATEGY

To be effective, developmental testing of the Land Warrior System and the helmet-mounted display must expand in scope and increase the human and operational variables introduced. The introduction of operational considerations should be executed in a controlled and prioritized manner with a deliberate research and testing sequence using controlled field experiments. To be effective and cost-effective, subsequent operational testing should be implemented at the small unit level, minimizing larger-scale testing scenarios. The testing should have increased scope, with longer durations or cycles of task performance than have been used in the past. Soldier in-the-loop testing with increased emphasis on the number of operational considerations should be an early objective.

In recent years, the Army has made major strides in developing, refining, and integrating constructive models, virtual simulations, and live simulations in the training arena. However, these advances have not been effectively applied in the area of test and evaluation, especially in the early stages of the development cycle. Rapidly growing modeling and simulation technologies appear to present a significant opportunity for enhancing the integration of human factors considerations into the development of the Land Warrior program. Simulation technology now presents a real possibility of assessing human performance. The U.S. Army Training and Doctrine Command is developing a series of virtual simulations and simulators that may be very useful in assessing human performance. However, there is insufficient attention to the measurement of human performance in these activities at the present time. Although the virtual simulations in their current form do not meet the design consideration needs for the helmet-mounted display per se, it is an example of simulation approaches that should be considered in planning and structuring the research and testing program for the helmet-mounted devices. As described by the user (U.S. Army Infantry School), the Land Warrior System is not being employed. This void must be filled prior to the development of a detailed test and evaluation program. Modeling and simulation can play useful role in specifying the parameters for use of the Land Warrior System use in the field.

By simulating a variety of threat environments, ranges of employment options, and different organizational structures, it should be possible to assess detection ranges, friendly and enemy engagement ranges, and the impact of a host of environmental and electronic warfare issues on the battlefield. The ability to quickly and relatively inexpensively vary potential battlefield conditions would allow the investigation of difficult performance specifications.

As suggested above, the design, testing, and evaluation processes of the Land Warrior program and the helmet-mounted display can benefit from expanding the capabilities and application of constructive and virtual simulations. Such approaches could reduce the amount of operational testing, but they will not entirely eliminate the need for live simulations and tactical exercises. Executing experiments and employment exercises with real soldiers and their leaders in the loop has always been a primary part of the Army's test and evaluation process. If, however, live simulations are to have an impact on the design of the helmet-mounted display, a more stringent scientific methodology must be adopted for early operational testing.

The large number of integrated subsystems that are part of the Land Warrior System make it difficult to measure the effects on human performance of a specific subsystem. The ability to measure the speed and to automatically locate and estimate the range of a potential enemy target, coupled with an automatic request for indirect fires, does not measure the soldier's concurrent potential to detect or failure to detect other, more significant threats that might mitigate the sheer speed of engaging one target. In order to properly evaluate the task-loading effects of the helmet-mounted display and the interfaces within the Land Warrior System, the discrete subsystems, sequences of events, and associated human task loads must be tested in a controlled environment. Such an approach will allow measurement of the addressed factors with a sampling of typical soldiers.

User-Centered Design

In the early stages of the research, development, test, and evaluation process, critical design decisions are often made by scientists and engineers who are knowledgeable concerning the technology but less expert in the operational employment of the new system. Without consistent and thorough collaboration with knowledgeable user representatives, the evolving design is driven by the engineer's and technician's view of what is important or what is possible, with less focus on what is most needed by the user. The Land Warrior System and the helmet-mounted display represent a major step forward in the integration of technology. To be effective, the research, development and design process must have more than user inputs; it must have user involvement and commitment. User-centered design is a significant aspect of the three-tier process described earlier.

User-centered design is an outcome-oriented approach to system development. This phrase means that user needs, capabilities, limitations, and preferences are not only taken into account when design decisions are being made, but also can actually drive such decisions. It also means that the user is the focus of attention throughout the development, and that the developing system is tested and evaluated in a realistic operational environment early in the development process. The objective is to build a high level of user acceptance and confidence in the system. The Land Warrior System will only be effective if infantry soldiers can achieve enthusiastic confidence in their abilities to perform their mission using it.

The Army development process is well defined. The role of the Training and Doctrine Command as the user representative is well known and accepted within the development community. However, the level of knowledgeable and focused user interaction in the design process is, all too often, lacking. Using the philosophy of user-centered design can aid the accelerated development cycle that has been outlined for the Land Warrior System and the helmet-mounted display.

Critical Test Issues

In order to make human factors a vital part of the test and evaluation process for the helmet-mounted display, it is necessary to identify the critical human factors research questions that must be answered during the test and evaluation process. After careful analysis, the panel identified the following test issues in four broad areas as critical to the design of displays for the individual soldier:

1. Local situation awareness,
2. Visual perception,
3. Cognitive workload, and
4. Soldier qualifications.

Local Situation Awareness

Do the technologies and capabilities embedded in the helmet-mounted display provide a significant improvement in a soldier's local situation awareness under the expected environmental and operational conditions? An operational way to phrase this would be, "Does a dismounted infantry soldier equipped with a helmet-mounted display have a significantly improved ability to detect, identify, and effectively engage enemy targets over a currently equipped soldier?" The discussion on situation awareness in Chapter 3 addressed both global and local situation awareness. From the soldier's point of view, both his survival and his performance enhancement involve local situation awareness. Under what set of environmental or tactical employment conditions does the helmet-mounted display enhance soldier response time? Clearly, on the basis of what we know from current human factors research and technology development, enhanced performance does not result simply by adding a helmet-mounted display capability. The criteria associated with this test issue deal with the employment conditions under which enhanced performance is sought. Performance comparisons of the currently equipped soldier with a Land Warrior-equipped soldier should be integral to the test plan.

Visual Perception

Does the helmet-mounted display significantly improve the visual perception of a soldier and his ability to perform operational tasks under tactical and operational conditions? This report has addressed some of the major visual perception issues that the design of the helmet-mounted display raises. More research is required to define operational parameters for task performance under likely operational conditions to establish acceptable trade-offs to maximize and enhance human visual capabilities.

Cognitive Workload

How does the cognitive workload created by the helmet-mounted display affect task performance by soldiers under tactical and operational conditions? Measuring cognitive workload is the most significant problem area that affects the design of the helmet-mounted display and the one that is most difficult to objectively quantify in testing. The outcomes of workload measurement are dependent on environmental conditions, individual human cognitive capabilities, and a large number of other variables. The research and testing program must first be focused on isolating the contribution of key variables under controlled conditions.

Soldier Qualifications

Can current infantry soldiers, with existing skill qualifications, perform their expected operational tasks more effectively with the helmet-mounted display and its designed capabilities? A basic assumption of the Land Warrior program is that the system will be used by basic infantry soldiers. This implies that no special selection criteria different from the current standards will be needed. Based on current research in human performance, it is unclear whether soldiers who meet the current Army criteria can operate the Land Warrior System and the complex helmet-mounted display and achieve enhanced performance under realistic battlefield conditions. The testing program for the helmet-mounted display should be designed to investigate the boundaries of effective performance for soldiers who meet the current standards and with soldiers from controlled groups who meet higher and more selective entrance requirements.

Cross-Cutting Issues

A number of topics relate to more than one test issue. One is the effects of stress on performance; another is the organization of information and the methods selected to display that information. Stress may be caused by physical or cognitive workload or by poorly fitting equipment. Such stresses can influence situation awareness and the ability of the soldier to process and use information. Specifically, although there is some knowledge of separate physical and cognitive workload issues, there is almost no research information on their combined effects. As stress increases, attention narrows to a few immediately obvious informational cues. Battle places enormous stress on the soldier, so one should expect such narrowing to occur. Likewise, there are many unresolved questions related to the stress imposed by poor and ill-fitting equipment. As the ergonomic portions of the program continue, developers need to be informed by the stress reactions of users; otherwise, apparent gains may well be wiped out under actual combat operations. What is needed is much more basic research on this issue, tied to applied research questions associated with stressful performance using the helmet-mounted display.

Questions of display design relate to all four test issues. One area in need of research attention is the hierarchical ordering of information from immediate threat to minor operational concerns and evaluating alternative presentation sequences and formats. Another research area concerns the allocation of information to visual versus auditory channels and the applicability of advanced technology, such as three-dimensional audio, to making these allocation decisions. Also, research is needed on the way in which graphic displays are structured and how these displays are formatted into standard iconic symbols for action. As discussed earlier, it is possible that new information-processing and display capabilities could be used to reduce stress by providing a global help function (e.g., location of nearest friendly force) at all times. A critical area of software development is the provision of this information in a secure manner. One could even imagine that an adaptive interface system could be used to on-load the soldier during periods of boredom and off-load the soldier during periods of high workload.

IMPLEMENTING THE STRATEGY

The panel proposes a three-tier concept of research and testing. The ideas of overlaying developmental and operational testing and expanding the use of simulations are not new to the Army, but embedding a level of controlled field testing with real users is. The Army test community has long recognized the potential of combining developmental and operational testing and evaluation as a cost- and time-saving approach. However, the relative difficulty of assessing technical performance to meet specific technical contract specifications, versus evaluating system performance in typical user's hands under operational conditions, has never been adequately addressed. The latter requires integration of operational performance criteria and user personnel into technical and developmental tests as well as controlled field tests. In doing so, the experimental design features used in behavioral research should be incorporated, including the procedures associated with sampling human subjects that specify the sample characteristics and size requirements needed to draw robust conclusions.

Such testing must be focused on critical performance criteria that can be specified in operational terms. This has been a stated objective within the Army's acquisition guidelines and the test and evaluation community for a number of years. The Land Warrior System Specification (Specification A3246133G) dated February 22, 1995, documents the requirements for the system and includes many operational criteria and parameters. However, the specifications are not at a level of operational specificity to provide insight into the human performance issues associated with the helmet-mounted display. Greater focus is needed on the types of human performance expected under realistic combat conditions. Table 7-1 highlights existing requirements for helmet-mounted display subsystems and the critical human-factors related test issues that must be resolved to permit the necessary specification of design requirements that address human operational performance criteria. In the panel's view, these issues, which relate directly to the broader test issues discussed previously, must drive the helmet-mounted display design and research and testing process. Although cognitive overload is addressed for each component in Table 7-1, it is important to keep in mind that questions of workload cannot be answered in a piecemeal fashion but require the integration of components.

A system's performance requirements should define the environmental conditions under which the system must operate. Human performance is influenced by the single or combined effects of numerous factors that enhance or degrade and/or facilitate or interfere with that performance, depending on the particular situation. These factors, referred to as performance-shaping factors, are aspects of the human-machine system that influence behavior and affect the time or accuracy of the human response (Swain and Guttman, 1980). To be relevant, performance-shaping factors must have the potential for significantly affecting the magnitude of, frequency of, or variability in human performance. Identifying performance-shaping factors and establishing their effect on human performance is an important aspect of the research and testing necessary for designing the helmet-mounted display.

Performance-shaping factors are grouped as operational, equipment, task, personnel, and environmental factors. Operational factors include doctrine and tactics, length of time of use, and system objective. Equipment performance-shaping factors for the helmet-mounted display include the physical parameters, operating characteristics, display layout, and reliability and maintainability. Personnel factors relate to items such as training, experience level of the user, motivation, skill level of the user, fatigue, monotony, attitude, and workload. Task factors such as complexity, duration, repetitiveness and environmental factors such as distracting stimuli, vibration, motion, visibility, temperature, and noise must all be addressed in the design of specific test iterations to build a better understanding of the impacts on human performance of use of the helmet-mounted display.

Research and Testing Approaches

Table 7-2 lists the key research issues identified by the panel and proposes a general approach to addressing them. Included are the subsystems of the HMD and the subsystems that interface with the HMD. The test conditions and test criteria listed are representative of the conditions and measures that are most significant. The intent of the table is to list the kinds of conditions that must be addressed to assess human performance differences and the types of standards or criteria that should be considered. As we've said, the intent of the proposed three-tier test approach is (1) to use infantry soldiers in the research during controlled laboratory and bench testing whenever practical as well as under controlled and operational field conditions and (2) to strictly control test variables in an attempt to isolate the human performance differences.

Identifying Measures of Effectiveness

A measure of effectiveness is a measure with a standard or criterion that allows testers or researchers to evaluate performance. A measure of performance is a quantitative indicator without a standard that describes performance but does not evaluate it. The thrust of the proposed research and testing effort for the helmet-mounted display should be directed toward assessing the human performance of soldiers against operational measures of effectiveness early in the development cycle. The intent is to use controlled experiments in a much more operational way and in smaller structured test scenarios. Although it would be desirable to build simulations that allow the conduct of many iterations using actual soldiers, controlled testing can be initiated now and does not have to be executed on a large scale to obtain significant levels of human performance data.

Tables 7-3 and 7-4 illustrate links among the Land Warrior System's mission tasks, potential measures of effectiveness and the proposed measures of effectiveness. The individual soldier tasks using the Land Warrior System are listed in Table 7-3 for a reconnaissance mission are not an exhaustive set. However, they represent tasks that we would specifically expect the individual soldier equipped with the Land Warrior System to perform in an enhanced manner over a currently equipped infantry soldier.

Table 7-4 presents a similar set of tasks for a squad leader equipped with the Land Warrior System linked to a series of potential measures of effectiveness.

TABLE 7-1 Land Warrior/Helmet-Mounted Display Capability and Critical Issues

Subsystem Description/Requirement	Critical Helmet-Mounted Display Test/Research Issue
GPS/computer radio subsystem - Provides position location data - Integrates information via software - Operates under all operational conditions	Does the GPS/computer radio subsystem cognitively overload the llB Land Warrior soldier? (#3)
Weapon subsystem - Integrates target acquisition, location, and marking - Increases speed of engagement - Operates under all operational conditions	Can Land Warrior soldiers more effectively engage targets with the integrated helmet-mounted display sight systems? (#l)
Computer processor in the computer radio subsystem - Executes software programs and computations - Processes sensor inputs - Operable by llB soldiers (ASVAB scores)	Do the functions of the computer processor cognitively overload the llB Land Warrior soldier? (#3) Can llBs who meet current qualification standards effectively employ the CPS? (#4)
Sensor/display assembly (night sensor display) - Displays clear video images - Operates under all operational conditions - Controls permit tactile identification - Adjustments while wearing gloves - Operable by llB soldiers (ASVAB scores)	Is local situational awareness of the Land Warrior soldier increased with the night sensor display? (#l) Does night sensor display enhance Land Warrior visual perception? (#2) Can llBs who meet current qualification standards effectively employ the night sensor display? (#4) Do the functions of the night sensor display cognitively overload the llB Land Warrior soldier? (#3)
Sensor/display assembly (day sensor display) - Displays clear video images - Operates under all operational conditions - Compatible with laser eye protection optical - Operable by llB soldiers (ASVAB scores)	Is local situational awareness of the Land Warrior soldier increased with the day sensor display? (#l) Does day sensor display enhance Land Warrior visual perception? (#2) Do the functions of the day sensor display cognitively overload the llB Land Warrior soldier? (#3)
Remote input pointing device - Operates under all operational conditions - Operable by llB soldiers (ASVAB scores) - Tactile feedback with gloves	Does the remote input pointing device support the workload of the Land Warrior soldier effectively? (#3) Can llBs who meet current qualification standards effectively employ the remote input pointing device? (#4)

TABLE 7-1 Continued

Subsystem Description/Requirement	Critical Helmet-Mounted Display Test/Research Issue
Audio amplifier of computer radio system - Allows mixing of computer and radio outputs - Controls overall audio volume level - Provides audio emphasis control	Is local situational awareness of the Land Warrior soldier increased by use of the audio amplifier? (#l)
Audio headset (microphone/speaker) - Receives/transmits radio and computer generated tones and outputs - Useable in airborne and air assault operations - Noise cancellation microphone	Is local situational awareness of the Land Warrior soldier increased by use of the audio headset? (#l) Do the functions of the audio headset cognitively overload the llB Land Warrior soldier? (#3)
Ballistic helmet assembly (shell and suspension) - Fits 5-95 percent male with improved stability - Standard PSGT tooling (40 percent improve protection)	Does the ballistic helmet assembly provide the required level of stability to support Land Warrior missions?
Hand-held display and mini keyboard - Operable while wearing gloves - Tactile feedback with gloves and NBC	Do the functions of the night sensor display cognitively overload the llB Land Warrior soldier? (#3)

The last column lists candidate measures of performance that would be used to quantify performance differences. Both tables are representative of the types of objective measures that need to be constructed for the individual soldier, squad leader, and platoon leader for all tactical infantry Land Warrior missions.

The potential measures of effectiveness and measures of performance illustrated in Tables 7-3 and 7-4 are basic operational measures that could be applied in all three tiers of the proposed research and testing process. The intent is to conduct controlled experiments with test and control groups using representative soldiers under operational conditions starting early in the research and testing cycle. Such an approach would facilitate human performance assessment, and at the same time, minimize the amount of follow on operational testing required to support the Army's acquisition process. The key is establishing a consistent set of agreed upon operational conditions that can be applied in all testing phases.

TABLE 7-2 Research and Testing Approaches for the Helmet-Mounted Display

Equipment	Design Research	Test Approach	Test Conditions	Test Criteria
Helmet-mounted display and weapon sight	Distance—offset of sensor/display	Controlled field experiments Operational test and evaluation	In buildings and with obstacles	Time to detect targets
Helmet-mounted display/night vision goggle and TWS	Resolution Field of view Chromaticity	Controlled laboratory/bench tests Controlled field experiments Operational test and evaluation	± Levels of visibility, distance, background, contrast, target-nontarget similarity	Time to acquire Time to identify targets Navigation time and error rate Obstacle error rate (in lab and operational field conditions)
Helmet-mounted display/night vision goggle and TWS	Monocular versus biocular versus binocular	Controlled laboratory/bench tests Controlled field experiments Operational test and evaluation	± Levels of employment, duration of use, light condition	Time to acquire Time to identify targets Navigation error rate Physiological effects Obstacle error rate Time to detect unaided after aided

Helmet-mounted display versus hand held device	Information to be communicated Display format Input format	Controlled field experiments Operational test and evaluation	± Levels of duration, fatigue, stressors, experience	Time to perform tasks Task error rate Level of local situational awareness
Helmet-mounted display/audio modality (monaural)	Information cues in display: Placement, format, types: attentions, warnings	Controlled field experiments	± Levels of visibility, distance, contrast, sensitivity, noise, task loading	Percent cues detected Time to detect cues Time to identify information in environment
Helmet-mounted display/3-D audio displays (biaural)	Resolution needed Types needed: Way-point navigational Localized communication Enhanced hearing Head tracker Stereo audio	Controlled laboratory/bench tests	LOS radio acoustically controlled; noise at defined volume/ frequency content; visual displays with control target	Percent cues detected Time to detect cues Time to identify information in environment
		Controlled field experiments	Outdoors, ± ambient noise; LOS, masked at min/max ranges; visual targets at given distances; day, dusk, night	Percent correct Time to detect target information Time to identify target

Continued on next page

TABLE 7-2 Continued

Equipment	Design Research	Test Approach	Test Conditions	Test Criteria
		Operational field testing	Field exercise, noise, fatigue, masking, varied ranges and times	Percent correct recognition of standard numbers and letters Percent correct recognition of standard message format content
Speech recognition (N.B.: Assessing system recognition of user/command voice input, converting to text)	Vocabulary design Speaker display versus speaker independent Discrete versus continuous	Controlled laboratory/bench tests	LOS radio acoustically controlled environment; insertion of noise at defined volume/frequency content	Percent correct words on modified rhyme test Percent correct of standard numbers and letters
		Controlled field experiments	Outdoors, ± ambient noise; LOS, masked at minimum/maximum ranges; visual targets at controlled distances	Percent correct and time for target information detection and identification
		Operational field testing	Field exercise, noise, fatigue, masking, varied ranges—day, dusk, night	

				Percent correct recognition of standard message format contents
				Recognition and application of message report contents
Soldier/radio	Intelligibility of speech	Controlled laboratory/bench tests	LOS radio acoustically controlled environment; insertion of noise at defined volume/ frequency content	Percent correct words on modified rhyme test
				Percent correct of standard numbers and letters
		Controlled field experiments	Outdoors, ± ambient noise; LOS, masked at minimum/maximum ranges; visual targets at controlled distances	Percent correct and time for target information detection and identification
		Operational test and evaluation	Field exercise, noise, fatigue, masking, varied ranges—day, dusk, night	Percent correct recognition of standard message format contents
				Recognition and application of message report contents

TABLE 7-3 Soldier Battle Tasks and Potential Measures of Effectiveness (MOE)/Measures of Performance (MOP)

Land Warrior Task	Potential MOE	Potential MOP
Receive squad order	Land Warrior soldiers have better knowledge of mission	Time required to issue order Percent of soldiers able to state mission and location
Move tactically	Moving Land Warrior soldiers are less detectable	Time required to execute movement Number enemy Land Warrior detections Detection range
Identify ORPs	Land Warrior soldiers correctly identify ORPs	Percent ORPs correctly identified
Observe/ listen for enemy	Land Warrior soldiers detect enemy activity more effectively	Percent enemy activity detected Number of correct detections
Conduct reconnaissance	Land Warrior soldiers are able to conduct accurate reconnaissance in less time	Time to conduct reconnaissance Ratio of enemy Land Warrior detections to Land Warrior enemy detections
Locate objective area	Land Warrior soldiers locate objects more effectively	Time to locate objectives Percent correct identifications

continued on next page

TABLE 7-3 Continued

Land Warrior Task	Potential MOE	Potential MOP
Observe enemy activity	Land Warrior soldiers detect enemy activity more effectively	Percent enemy activity detected
		Number of correct detections
Record information	Reconnaissance results are more complete	Time required to record results
		Percent observations recorded
Report enemy information	Land Warrior reports are more timely and accurate.	Time required to deliver report
		Number of reports received
		Percent complete reports

CONCLUSION

The complex human performance issues associated with the Land Warrior System require a rigorous program of research integrated with effective user-oriented field experiments and operational tests. Applying advanced technologies to enhance the performance of infantry soldiers cannot be achieved without also enhancing individual cognitive processing and individual local situation awareness under expected future battlefield conditions. To successfully achieve the optimistic capabilities envisioned in the Land Warrior program, there are significant elements of developmental risk that must be minimized by further research and controlled testing. The research must be completed in time to help drive the design of the helmet-mounted display.

TABLE 7-4 Squad Leader Battle Tasks and Potential Measures of Effectiveness (MOE)/Measures of Performance (MOP)

Reconnaissance Task	Potential MOE	Potential MOP
Receive platoon order	Land Warrior squad leaders have more order complete mission knowledge	Time to receive platoon
		Percent squad leaders able to correctly state 5 Para Order details.
Evaluate move route	Land Warrior squad leaders are more effective in route evaluations	Time to complete route evaluation Percent identification of critical features
Move tactically	Moving Land Warrior squads are less detectable	Time to execute movement Number enemy Land Warrior detections Detection range
Control squad movement	Squad movement is more effective	Time to execute movement Average squad dispersion
Land navigate	Squad leaders navigate more accurately	Time to move and identify ORPs Percent correct location reports
Identify ORPs	Squads correctly identify ORPs	Percent ORPs correctly identified
Determine location	Squad locations are accurately known	Percent time squad leader correctly locates elements
Report location	Squad locations are accurately reported	Percent time higher command levels know the correct squad location
Observe/listen for enemy	Squads detect enemy	Percent enemy activity more effectively detected Number correct detections

continued on next page

TABLE 7-4 Continued

Reconnaissance Task	Potential MOE	Potential MOP
Conduct reconnaissance	Squads conduct effective reconnaissances	Time to conduct reconnaissance Percent enemy detections to Land Warrior detections
Locate objective area	Squads locate objectives effectively	Time to locate objectives Percent correct identifications
Observe enemy activity	Squads detect enemy more effectively	Percent enemy activity detected Number correct detections
Record information	Squad reconnaissance results are more complete	Time required to record results Percent observations recorded
Receive enemy information reports	Squad reports of enemy are accurate and timely	Percent enemy activity reported Number correct reports Time required to record results Percent observations recorded
Report enemy information	Squad reports of enemy are more timely and accurate	Percent enemy activity reported Number correct reports Time required to record results Percent observations recorded
Report situation	Squad reports are more timely and accurate	Time required to deliver report Number enemy reports received Percent complete reports

8

Recommendations

In this chapter we present the conclusions and recommendations given the highest priority by the panel. Each chapter also contains a set of conclusions and design guidelines.

MAJOR RECOMMENDATIONS

The panel's overarching conclusion, after reviewing the available Land Warrior specifications and the existing human factors research findings that apply to those specifications, is that the proposed monochrome, low-resolution monocular as compared to a binocular display will, in most situations, degrade performance in the field and may have unacceptable implications for training and selection.

> RECOMMENDATION 1: The Army should proceed in an experimental mode, comparing the positive and negative performance implications of the monocular helmet-mounted display with alternative technologies. One fruitful approach would be to select alternative promising technologies, including hand-held and other types of displays, issue them to experimental groups, and compare performance.

The challenges associated with the helmet-mounted display for the dismounted infantry soldier are considerable, and the body of human factors knowledge about the relative merits of the numerous design options is, at present, limited. Although a substantial amount of work on such displays in the aviation environment has been done, the direct transfer of results to the infantry environment is not possible due to differences in task conditions and performance requirements. A major difference is that pilots are relatively stable (not moving) on a moving platform, whereas the infantry soldier is moving on a fixed platform (the ground). With regard to specific elements of the helmet-mounted display technology proposed by the Army, the panel makes the following conclusions.

After careful review of the available data on the human visual system, the panel concludes that a monochrome, low-resolution, monocular display will, in most situations, degrade human performance compared with a binocular system. The chosen technical approach offers less than optimal sensor resolution, less than optimal spatial and temporal display resolution, and less than ideal field of view, contrast, and chromaticity. These factors, coupled with absent or anomalous stereoscopic visual depth information, tend to keep the human accommodation and vergence systems running open-loop and drastically degrade or even eliminate visual depth information. The result can be eyestrain, fatigue, and possibly spatial disorientation, as well as loss of equilibrium and loss of form and layout perception. One important area of investigation is the potential long-term consequences of monocular viewing and rivalry.

The panel further concludes that, even if the visual perception issues associated with the proposed helmet-mounted display technology are resolved, shifting the infantry soldier's attention away from the battlefield toward a computer-generated display raises other critical issues. The helmet-mounted display may compromise local situation awareness (location, presence of enemies, terrain and object perception) by competing for mental resources and affecting perceptual processes. Also, the potential increase in cognitive workload associated with processing and applying the information may be in itself prohibitive.

> RECOMMENDATION 2: If the display of digital data partially occludes the soldier's view of the environment, then hand-held or wrist-mounted displays should be considered as an alternative to the helmet-mounted display for digital data in order to reduce the likelihood of negatively affecting the soldier's local situation awareness.

RECOMMENDED RESEARCH AGENDA

The proposed Land Warrior System can be a valuable research tool, if the Army takes an experimental approach to its development. If put into the hands of users in a experimental mode, the Army can establish baseline data and threshold values for future developmental efforts.[1]

---

[1]A 1996 report from the General Accounting Office concludes that several human factors issues associated with the Land Warrior System are not yet resolved (U.S. General Accounting Office, 1996). We concur with the general thrust of the report, although our view is that sufficient specimens of the Land Warrior System—including the helmet-mounted display subsystem—should be acquired for research purposes and compared with alternative technologies. Evaluating such specimens in a realistic setting should help answer the questions raised by the GAO report as well as those raised in this committee's report.

1. Research should be undertaken about the relationship among design attributes, human attributes, and successful performance for the Land Warrior System. Significant increases or changes in soldier skills and abilities may be required as a consequence of these technologies. Effective personnel selection and training for the system will depend on understanding these relationships.

2. Threshold values are needed for screen clutter, gray scale, limits of spatial and temporal resolution, the impact of visual acuity differences in soldiers, short-term memory limits in processing the information, individual susceptibility to various levels of incapacitation associated with visual rivalry, depth cues, field of view (versus resolution) values, delivery modality preferences and trade-offs, and the impact of attentional narrowing.

3. Although a viewer can compensate to some extent for a small field of view by making more head movements to obtain a series of small glimpses of the environment and can compensate for the loss of stereopsis by using small head movements to provide depth information through parallax, the significant optical persistence characteristics proposed for the helmet-mounted display make such corrections problematic. Furthermore, there is currently no accepted cognitive theory from which one can set the bounds of one glimpse. Important research questions include: How are successive glimpses of the display organized by the visual system into a single perceptual image? How much structural overlap is required? Over how much delay? Over how many shifts in view? How is this information combined with outside information? The overall impact of the system on soldier situation awareness, when used in combination with other equipment and information in the battlefield environment, must be addressed.

4. The provision for remaining in a protected position while extending, aiming, and firing the rifle is laudable. However, the question of rifle stabilization has not been adequately addressed. In the current weapon system, the sighting device may be considered accurate but the aiming variance associated with holding the unsupported weapon stable is large, particularly under conditions of sustained performance. The absence of image stabilization associated with a helmet-mounted display is also an issue for viewing and sighting. Even if one assumes the design of a new helmet, skull skin movement can range from 0.5 to 1.5 inches. Tests of the accuracy of the aiming of the rifle should be included in the field research program.

5. Physical factors in the battlefield—such as heat, cold, vibration, and noise—all have implications for the design of helmet-mounted displays and the information they provide to soldiers. Research is needed to explore the relationship between stress on the electronic battlefield and performance. One important area is the potential effects of vibration and small shifts in helmet alignment (caused by walking or more violent motion) on the effective use of a helmet-mounted display. Another is the combined effects of physical and mental workload on soldier performance over extended periods of time.

6. The low-resolution, monochrome, monocular display presents challenges associated with movement, target recognition and identification, hit accuracy, map reading speed, cognitive workload, and functionality. The benefits claimed for the Land Warrior System involve the use of the unaided eye at night as a baseline for comparison—not the unaided eye during the day or with night vision goggles. Comparisons of the Land Warrior monocular system should be made with the existing biocular night vision goggles worn by infantry soldiers and the binocular night vision goggles worn by aviators. Comparisons made by researchers at Aberdeen Proving Ground (CuQlock-Knopp et al., 1994) suggest that people navigate their environment better at night with binocular viewing (compared with monocular viewing).

7. In a given battle situation, the pace of engagement would allow only seconds for reading a display. Presenting messages to the ear may be a more effective way to communicate with the engaged soldier. What data are critical to show visually to the infantry soldier during combat must be investigated. The minimum and maximum threshold values of visually displayed data need to be determined. Research is needed on the perceptual narrowing threshold between audio and visual displays. The trade-offs among free field attenuation devices, active cancellation, normal communications traffic, signals, icons, and working memory capacity need research.

8. Helmet weight and the distribution of that weight have the potential to degrade the performance of the infantry soldier. Currently there is no evidence demonstrating the effects of the weight requirement specified for the proposed helmet-mounted display in the Land Warrior System. Likewise, no studies have been done to ensure that freedom of head movement is not impaired and that helmet stability is congruent with aiming a hand-held weapon. As part of the larger program of applied field research, testing and evaluation should be undertaken to ensure that the weight and distribution of the equipment does not interfere with the ability of the soldier to move freely and aim his weapon accurately.

RECOMMENDED DESIGN GUIDELINES

We recommend that the following design guidelines be adopted to maximize the soldier's situation awareness and facilitate his ability to process information efficiently:

1. The helmet-mounted display should minimize the degree to which it is a physical barrier to acquiring information about the world (e.g., occludes or alters normal hearing and vision). It should enhance regular sensory input only when needed (e.g., targeting support, night vision).

2. The helmet-mounted display should minimize both attentional distraction and the cognitive load it places on the user by providing integrated information in

task-oriented sequence, by reducing extraneous information, and by minimizing memory requirements.

3. The helmet-mounted display should provide salient cueing—directing the soldier's attention to important information. The audio mode is generally preferable for simple cueing and the transmission of time-critical information.

4. Spatial, topographic, and positional information should be presented by graphics that have been well learned by soldiers (e.g., standardized map symbols) in order to facilitate rapid and accurate interpretation.

5. Whenever possible, the system should simplify the presentation of data entry and system control options. Simplification of these tasks will minimize workload in high-load situations, such as battlefield engagements.

References

EXECUTIVE SUMMARY

Franks, General F.M., Jr.
1994 Review of events of the Gulf war. *Vital Speeches of the Day* 60(15):453-458 (May 15, 1994).

U.S. Department of the Army
1993 Mission Need Statement for the Enhanced Integrated Soldier System-Dismounted (MNS). September 8, 1993 Memorandum for Assistant Secretary of the Army (Research, Development and Acquisition). Documentation No. DAMO-FDZ. Washington, DC: Army Pentagon.
1994 Operational Requirements Document (ORD) for the Land Warrior (LW) Systems. April 13, 1994 Memorandum, Documentation No. ATCD-SE (70-1f). Washington, DC: Army Pentagon.

U.S. General Accounting Office
1996 *Battlefield Automation* (GAO/NSAID-96-190). Washington, DC: U.S. Government Printing Office.

CHAPTER 1

Adkins, R., W. Murphy, M. Hemenway, R. Archer, and L. Bayless
1995 Hardman III Analysis of the Land Warrior System. Army Research Laboratory Technical Report ARL-CR-1995. Dynamics Research Corporation: Andover, MA.

Buck, J.R., and B.H. Kantrowitz
1983 Decision Making. Pp. 428 ff in B.H. Kantrowitz, and R.D. Sorkin, eds., *Human Factors: Understanding People-System Relationships*. New York: Wiley.

Clark, Bruce C., C.G.
1959 Speech made at the Fourth Annual Human Factors Meeting.

Guilmartin, John F., and D.W. Jacobwitz, eds.
1984 Technology, Primary Group Cohesion, and Tactics as Determinants of Success in Weapons System Design: A Historical Analysis of an Interactive Process. U.S. Air Force Command and Staff College.
U.S. Department of the Army
1992 *Field Manual (FM) 7-8 Infantry Rifle Platoon and Squad.* April 22. Washington, DC: Army Pentagon.
1994 Force XXI—America's Army for the 21st Century. *Army Focus*, September.
U.S. Department of the Army, Training and Command Center
1994 *Force XXI Operations, A Concept for the Evolution of Full-Dimensional Operations for the Strategic Army of the Early Twenty-First Century.* TRADOC Pamphlet No. 525-5. Washington, DC: Army Pentagon.

CHAPTER 2

Bowman, W., R. Little, and G. Sicilia, eds.
1986 *All Volunteer Force After a Decade.* Pergamon-Brassey International Defense Publishers.
Fleishman, E.
1975 Toward a taxonomy of human performance. *American Psychologist* 30:1127-1149.
Gordon, C.C., T. Churchill, C.E. Clauser, B. Bradtmiller, J.T. McConville, I. Tebbetts, and R.A. Walker
1988 *Anthropometric Survey of U.S. Army Personnel: Summary Statistics Interim Report.* Natick, MA: U.S. Army Natick Research, Development, and Engineering Center.
Green, B.F., Jr., and A.S. Mavor
1994 *Modeling Cost and Performance for Military Enlistment: Report of a Workshop.* Committee on Military Enlistment Standards, Commission on Behavioral and Social Sciences and Education, National Research Council. Washington, DC: National Academy Press.
Lehnus, Jerome D.
1994 *Shifting Perspectives of Military Service: The Views of Those Who Might Join.* Los Angeles, CA: American Psychological Association, Division 19.
Marshall, S.L.A.
1947 *Men Against Fire: The Problem of Battle Command in Future War.* New York: William Morrow and Company.
Salter, M.S.
1993 *Soldiers Integrated Protective Ensemble: The Soldier's Perspective.* Research report 1640. Alexandria, VA: U.S. Army Research Institute for the Behavioral and Social Sciences.
Wigdor, Alexandra K., and Bert F. Green, Jr., eds.
1991 *Performance Assessment for the Workplace.* Committee on the Performance of Military Personnel, Commission on Behavioral and Social Sciences and Education, National Research Council. Washington, DC: National Academy Press.
U.S. Department of the Army Natick RD&E Center
1994 *Target Audience Description.* Natick, MA: U.S. Army Natick Research, Development, and Engineering Center.
U.S. Department of the Army Research Institute
1988 *Implementing Embedded Training (ET): Volumes 1-10.* Arlington, VA.
U.S. Department of the Army
1982 U.S. Army Combined Arms Combat Development Activity. In *Soldier Machine Interface Requirements ("Complexity") Study*, May.

1992 *Field Manual (FM) 7-8 Infantry Rifle Platoon and Squad.* April 22, Washington, DC: Army Pentagon.

1994a Army Regulation 611-201 Enlisted Career Management Fields and Military Occupational Specialty. July 1. Washington, DC: Army Pentagon.

1994b Operational Requirements Document (ORD) for the Land Warrior (LW) Systems. April 13, 1994 Memorandum, Documentation No. ATCD-SE (70-1f). Washington, DC: Army Pentagon.

Witmer, B.G., and B.W. Knerr

1991 *A Guide for Early Embedded Training Decisions.* (Research Product 91-14). Alexandria, VA: U.S. Army Research Institute for the Behavioral and Social Sciences.

Zubal, O., N.W. Doss, J.A. Faughn, T.W. Haduch, and J.L. Thompson

1990 Human Engineering Laboratory Artillery P2NBC2 Prolonged Operations Demonstration. Technical Note 4-90. Aberdeen Proving Ground, MD: U.S. Army Human Engineering Laboratory.

Zubal, O., N.W. Doss, J.A. Faughn, K.P. Kysor, P.C. Szlyk, D.M. Caretti and I.V. Sils

1993 Evaluation of the Effect of Protective Clothing on the Ability of Self-Propelled Artillery (M109) Crews to Conduct and Sustain High Firing Rate Missions. Report number ARL-MR-98. Aberdeen Proving Ground, MD: U.S. Army Research Laboratory.

CHAPTER 3

Bacon, S.J.

1974 Arousal and the range of cue utilization. *Journal of Experimental Psychology*, 102:81-87.

Billings, C.E.

1991 *Human-Centered Aircraft Automation: A Concept and Guidelines* (NASA Technical Memorandum 103885). Moffett Field, CA: NASA Ames Research Center.

Broadbent, D.E.

1971 *Decision and Stress*. London: Academic Press.

Dominguez, C.

1994 Can SA be defined? Pp. 5-15 M. Vidulich, C. Dominguez, E. Vogel, and G. McMillan, eds., *Situation Awareness: Papers and Annotated Bibliography.* (AL/CF-TR-1994-0085). Wright-Patterson AFB, OH: Armstrong Medical Research Laboratory.

Endsley, M.R.

1987 *SAGAT: A Methodology for the Measurement of Situation Awareness*. NOR DOC 87-83. Hawthorne, CA: Northrop Corporation.

1988a Design and evaluation for situation awareness enhancement. Pp. 97-101 in *Proceedings of the Human Factors Society 32nd Annual Meeting*. Santa Monica, CA: Human Factors Society.

1988b Situation awareness Global Assessment Technique (SAGAT). Pp. 789-795 in *Proceedings of the National Aerospace and Electronics Conference (NAECON).* New York: IEEE.

1990a Predictive utility of an objective measure of situation awareness. Pp. 41-45 in *Proceedings of the Human Factors Society 34th Annual Meeting.* Santa Monica, CA: Human Factors Society.

1990b Situation Awareness in Dynamic Human Decision Making: Theory and Measurement. Unpublished doctoral dissertation, University of Southern California, Los Angeles, CA.

1993 Situation awareness and workload: Flip sides of the same coin. Pp. 906-911 in R.S. Jensen and D. Neumeister, eds, *Proceedings of the Seventh International Symposium on Aviation Psychology*. Columbus, OH: Department of Aviation, The Ohio State University.

1994 Situation awareness in dynamic human decision making: Theory. Pp. 27-58 in R.D. Gilson, D.J. Garland, and J.M. Koonce eds., *Situational Awareness in Complex Systems.* Daytona Beach, FL: Embry-Riddle Aeronautical University Press.

1995 Toward a theory of situation awareness. *Human Factors* 37(1):32-64.

1996 Situation awareness measurement in test and evaluation. Pp. 159-180 in T.G. O'Brien and S. G. Charlton, eds., *Handbook of Human Factors Testing and Evaluation.* Hillsdale, NJ: Lawrence Erlbaum.

Endsley, M.R., and C.A. Bolstad

1994 Individual differences in pilot situation awareness. *International Journal of Aviation Psychology* 4(3):241-264.

Endsley, M.R., and E.O. Kiris

1995 The out-of-the-loop performance problem and level of control in automation. *Human Factors* 37(2):381-394.

Fracker, M.L.

1988 A theory of situation assessment: Implications for measuring situation awareness. Pp. 102-106 in *Proceedings of the Human Factors Society 32nd Annual Meeting.* Santa Monica, CA: Human Factors Society.

Hockey, G.R.J.

1986 Changes in operator efficiency as a function of environmental stress, fatigue and circadian rhythms. Pp. 44/1-44/49 in K. Boff, L. Kaufman, and J. Thomas eds., *Handbook of Perception and Performance.* New York: John Wiley.

Janis, I.L.

1982 Decision making under stress. Pp. 69-87 in L. Goldberger and S. Breznitz, eds., *Handbook of Stress: Theoretical and Clinical Aspects.* New York: The Free Press.

Keinan, G.

1987 Decision making under stress: Scanning of alternatives under controllable and uncontrollable threats. *Journal of Personality and Social Psychology* 52(3):639-644.

Keinan, G., and N. Friedland

1987 Decision making under stress: Scanning of alternatives under physical threat. *Acta Psychologica* 64:219-228.

Kibbe, M.

1988 Information transfer from intelligent EW displays. Pp. 107-110 in *Proceedings of the Human Factors Society 32nd Annual Meeting.* Santa Monica, CA: Human Factors Society.

Klein, G.A.

1989 Recognition-primed decisions. Pp. 47-92 in W.B. Rouse, ed., *Advances in Man-Machine Systems Research.* Greenwich, CT: JAI Press, Inc.

Mandler, G.

1979 Thought processes, consciousness and stress. In V. Hamilton and D. M. Warburton, eds., *Human Stress and Cognition: An Information-Processing Approach.* Chichester: Wiley and Sons.

Moray, N.

1986 Monitoring behavior and supervisory control. Pp. 40/1-40/51 in K. Boff, ed., *Handbook of Perception and Human Performance.* New York: Wiley.

Nisbett, R.E., and T.D. Wilson

1977 Telling more than we can know: Verbal reports on mental processes. *Psychological Review* 84(3):231-259.

Sharit, J., and G. Salvendy

1982 Occupational stress: Review and reappraisal. *Human Factors* 24(2):129-162.

Taylor, R.M.
1990 Situational awareness rating technique (SART): The development of a tool for aircrew systems design. Pp. 3/1-3/17 in *Situational Awareness in Aerospace Operations (AGARD-CP-478).* Neuilly Sur Seine, France: NATO-AGARD.
Vidulich, M.A., and E.R. Hughes
1995 Testing a subjective metric of situation awareness. Pp. 1307-1311 in *Proceedings of the Human Factors Society 35th Annual Meeting*, (Vol. 2). Santa Monica, CA: Human Factors Society.
Wachtel, P.L.
1967 Conceptions of broad and narrow attention. *Psychological Bulletin* 68:417-429.
Weltman, G., J.E. Smith, and G.H. Egstrom
1971 Perceptual narrowing during simulated pressure-chamber exposure. *Human Factors* 13:99-107.
Wickens, C.D.
1992 *Engineering Psychology and Human Performance* (2nd ed.). Scranton, PA: Harper Collins.
Wickens, C.D., S. Todd, and K. Seidler
1989 Three-Dimensional Displays: Perception, Implementation, and Applications. Technical report, Aviation Research Laboratory, Institute of Aviation, University of Illinois at Urbana-Champaign.
Wiener, E.L., and R.E. Curry
1980 Flight deck automation: Promises and problems. *Ergonomics* 23(10):995-1011.
Wright, P.
1974 The harassed decision maker: Time pressures, distractions, and the use of evidence. *Journal of Applied Psychology* 59(5):555-561.

CHAPTER 4

Alpern, M.
1971 Effector mechanisms in vision. Pp. 369-394 in J.W. Kling and L.A. Riggs, eds., *Woodward and Schlosberg's Experimental Psychology.* New York: Academic Press.
Bachman, T., and J. Alik
1976 Integration and interruption in the masking of form by form. *Perception* 5:79-97.
Bemis, S.V., J.L. Leeds, and E.A. Winer
1988 Operator performance as a function of types of display: Conventional versus perspective. *Human Factors* 30(2):163-169.
Bennett, C.T., K.A. O'Donnell, and W.W. Johnson
1988 Dynamic perspective displays and the control of tilt-rotor aircraft in simulated flight. *Proceedings of the American Helicopter Society.* Washington, DC.
Biederman, I.
1981 On the semantics of a glance at a scene. Pp. 213-253 in M. Kubovy and J. Pomerantz, eds., *Perceptual Organization.* Hillsdale, NJ: Erlbaum.
1985 Human image understanding: Recent research and a theory. *Computer Vision, Graphics, and Imaging Processing* 32:29-73.
1995 Visual object recognition. Pp. 121-165 in S.F. Kosslyn and D.N. Osherson, eds., *An Invitation to Cognitive Sciences.* Cambridge, MA: MIT Press.
Boff, K.R., and J.E. Lincoln, eds.
1988 *Engineering Data Compendium.* Wright-Patterson Air Force Base, OH: Armstrong Medical Research Laboratory.

Brickner, M.S.
1989 Helicopter Flights with Night Vision Goggles: Human Factors Aspects. NASA Technical Memorandum 101039.

Brown, D.L., M.A. Berkley, and K.R. Jones
1978 Psychophysical changes in adult human visual capacities following occlusion of one eye. *Investigative Ophthalmology and Visual Science* 17 (Supplement):113.

Burnett, M., and W. Barfield
1991 An Evaluation of a Plan-View Versus Perspective Display for an Air Traffic Controller Task. Sixth International Symposium on Aviation Psychology. September, Columbus, OH.

Chapanis, A.S., and R.A. McCleary
1953 Interposition as a cue for the perception of relative distance. *Journal of Psychology* 48:113-132.

Crowley, J.S.
1991 *Human Factors of Night Vision Devices: Anecdotes from the Field Concerning Visual Illusions and Other Effects.* Report No. 91-15. Ft. Rucker, AL: U.S. Army Aeromedical Research Laboratory.

CuQlock-Knopp, V.G., W. Torgerson, D.E. Sipes, E. Bender, and J.O. Merritt
1994 Comparison of Monocular, Biocular, and Binocular Night Vision Goggles for Traversing Off-Road Terrain on Foot. ARL-TR-747, March. Aberdeen Proving Grounds, MD: U.S. Army Research Laboratory.

Cutting, J.E., and P.M. Vishton
1995 Perceiving layout and knowing distances: The integration, relative potency, and contextual use of different information about depth. In W. Epstein and S.J. Rogers, eds., *Handbook of Perception and Cognition: Volume 5, Perception of Space and Motion.* New York: Academic Press.

Dinnerstein, A.J., and M. Wertheimer
1957 Some determinants of phenomenal overlapping. *American Journal of Psychology* 70:21-37.

Ebenholtz, S.M.
1992 Motion sickness and oculomotor systems in virtual environment. *Presence Teleoperators and Virtual Environments* 1(1):3.
1988 *Sources of Asthenopia in Navy Flight Simulators.* Defense Technical Information Center Accession Report No. AD-A212 699. Ft. Rucker, AL: U.S. Army Aeromedical Research Laboratory.

Edgar, G.K., J.C.D. Pope, and I.R. Craig
1995 Visual accommodation problems with head-up and helmet-mounted displays? *Displays* 15:68-75.

Ellis, S., G. Tharp, and S. Smith
1991 Exocentric judgments in real environments and stereoscopic displays. Pp. 1442-1446 in *Proceedings of the Human Factors Society 35th Annual Meeting.* Santa Monica, CA: Human Factors Society.

Farrell, R.J., and J.M. Booth
1984 *Design Handbook for Imagery Interpretation Equipment.* Seattle, WA: Boeing Aerospace Company.

Fischer, E., R.F. Haines, and T. Price
1980 *Head-up Transition Behavior of Pilots with and without Head-up Display in Simulated Low-visibility Approaches.* Washington, DC: National Aeronautics and Space Administration.

Foley, J.M.
1987 Stereoscopic distance perception. Pp. 3.2-3.9 in S.R. Ellis, M.K. Kaiser, and A. Grunwald, eds., *Spatial Displays and Spatial Instruments*. NASA Conference Publication 10032. Moffett Field, CA: NASA Ames Research Laboratory.

Foyle, D.C., and M.K. Kaiser
1991 Pilot Distance Estimation with Unaided Vision, Night-Vision Goggles and Infrared Imagery. Digest of Technical Papers, SID International Symposium XXII. Anaheim, CA.

Gillam, B.
1995 The perception of spatial layout. In W. Epstein and S.J. Rogers, eds., *Handbook of Perception and Cognition: Volume 5, Perception of Space and Motion*. New York: Academic Press.

Grandjean, E.
1987 Design of VDT workstations. Pp. 1359-1297 in G. Salvendy, ed., *Handbook of Human Factors*. New York: Wiley.

Hart, S.G., and M.S. Brickner
1987 Helmet-mounted pilot night vision systems: Human factors issues. Pp. 13.1-13.21 in S.R. Ellis, M.K. Kaiser, and A. Grunwald, eds., *Spatial Displays and Spatial Instruments*. NASA Conference Publication number 10032. Moffett Field, CA: NASA Ames Research Center.

Helander, M.G., and B.A. Rupp
1984 An overview of standards and guidelines for visual display terminals. *Applied Ergonomics* 14:185-195.

Held, R., J. Dichgans, and J. Bauer
1975 Characteristics of moving visual scenes influencing spatial orientation. *Vision Research* 15:357-365.

Hochberg, J.
1980 Pictorial functions and perceptual structures. Pp. 47-93 in M. Hagen, ed., *The Perception of Pictures, 2*. New York: Academic Press.
1986 Representation of motion and space in video and cinematic displays. Pp. 1-22 in K. Boff, J. Thomas, and L. Kaufman, eds., *Handbook of Perception and Human Performance Volume 1*. New York: Wiley.

Hochberg, J., and V. Brooks
1996 The perception of motion pictures. Pp. 205-292 in M.P. Fieldman and E.C. Carterette, eds., *Cognitive Ecology*. San Diego, CA: Academic Press.

Hoffman, J.E., and S. Mueller
1994 An In-Depth Look at Selective Attention. Paper presented to 35th meeting of the Psychonomic Society, November. St. Louis, MO.

Howard, I.P.
1959 Some new subjective phenomena apparently due to interocular transfer. *Nature* 184:1516-1517.

Howard, I.P., and B.J. Rogers
1995 *Binocular Vision and Stereopsis*. Oxford, England: Oxford University Press.

Hughes, H.C., G. Nozawa, and F. Kitterle
1996 Global precedence, spatial frequency channels, and the statistics of natural images. *Journal of Cognitive Neuroscience* 8:197-230.

Human Factors Society
1988 *American National Standard for Human Factors Engineering of Visual Display Terminal Workstations*. ANSI/HFS:100-1988. Santa Monica, CA: Human Factors Society.

Iavecchia, Joyce H., Helene P. Iavecchia, and Stanley N. Roscoe
1988 Eye accommodation to head-up virtual images. *Human Factors* 30(6):689-702.

Johansson, B. and E. Börjesson

1990 Experiments on the optic sphere theory: Slant perception from central stimulation. Uppsala Psychological Reports Number 424. Uppsala University Department of Psychology.

Johnson, J.

1960 Analytical Description of Night Vision Devices. Pp. 177-200 in L. Biberman, ed., *Proceedings of the Seminar on Direct-Viewing Electro-Optical Aids to Night Vision.*

Julesz, B.

1971 *Foundations of Cyclopean Perception.* Chicago: University of Chicago Press.

Leibowitz, H.W., R.B. Post, T.H. Brandt, and J. Dichgans

1982 Implications of recent developments in dynamic spatial orientations and visual resolutions for vehicle guidance. Pp. 231-260 in A. Werthelm, W. Wagenaar, and H.W. Leibowitz, eds., *Tutorials in Motion Perception.* New York: Plenum Press.

Matin, L., and W. Li

1992 Visually perceived eye level: Changes induced by a pitched-from-vertical 2-line visual field. *Perception and Psychophysics* 18:257-289.

McAnn, R.S., D.C. Foyle, and J.C. Johnston

1992 Attentional limitations with head-up displays. To appear in *SID International Symposium Digest of Technical Papers*, XXII, 1991.

McCauley, Michael E., ed.

1984 *Research Issues in Simulator Sickness: Proceedings of a Workshop.* Committee on Human Factors, Commission on Behavioral and Social Sciences and Education, National Research Council. Washington, DC: National Academy Press.

McGreevy, M.W., and S.R. Ellis

1986 The effect of perspective geometry on judged spatial information instruments. *Human Factors* 28:439-456.

Neisser, U., and R. Becklin

1975 Selective looking: Attention to visually specified events. *Cognitive Psychology* 7:480-494.

O'Donnell, K.A., W.W. Johnson, C.T. Bennett, and A.V. Phatak

1988 The effect of perspective displays on altitude and stability control in simulated rotary wing flight. *Proceedings of the AIAA Flight Simulation Technologies Conference.* Atlanta, GA.

Ogle, K.N.

1953 On the problem of an international nomenclature for designating visual acuity. *American Journal of Ophthalmology* 37:909-921.

O'Kane, B.L.

1995 Validation of prediction models for target acquisition with electro-optical sensors. Pp. 192-218 in E. Peli, ed., *Vision Models of Target Detection and Recognition: In Memory of Arthur Menendez.* Singapore: World Scientific Publishing Co.

Peli, E.

1995 *Vision Models of Target Detection and Recognition: In Memory of Arthur Menendez.* Singapore: World Scientific Publishing Co.

Peterson, M.A., E.M.H. Harvey, and H.L. Weidenbacher

1991 Recognition inputs to figure-ground organization: Which route counts? *Journal of Experimental Psychology: Human Perception and Performance* 17:1075-1089.

Predebon, J.

1992 The role of instructions and familiar size in absolute judgments of size and distance. *Perception and Psychophysics* 51:344-354.

Previc, F.H.

1996 Vertical biases in perception and oculomotor control: Theory, data, and human factors applications. *Insight* 18:1-3.

Proffitt, D.R., B. Mukul, R. Grossweiler, and J. Midgett
1995 Perceiving geographical slant. *Psychonomic Bulletin and Review* 2:409-428.
RCA
1968 *Electro-Optics Handbook.* Harrison, NJ: RCA.
Riggs, L.A.
1971 Vision. Pp. 273-314 in J.W. Kling and L.A. Riggs, eds., *Experimental Psychology.* New York: Academic Press.
Rubin, N., K. Nakayama, and R. Shapley
1996 Enhanced perception of illusory contours in the lower versus upper visual hemifields. *Science* 271:651-653.
Rush, C.E., R.W. Verona, and J.S. Crowley
1990 *Human Factors and Safety Considerations of Night Vision Systems Flight Using Thermal Imaging Systems.* Report No. 90-10. Ft. Rucker, AL: U.S. Army Aeromedical Research Laboratory.
Schneider, G.E.
1969 Two visual systems. *Science* 163:895-902.
Schor, C.M.
1987 Spatial constraints of stereopsis in video displays. Pp. 2.1-2.13 in S.R. Ellis, M.K. Kaiser, and A. Grunwald, eds., *Spatial Displays and Spatial Instruments.* NASA Conference Publication 10032. Washington, DC: U.S. National Aeronautic and Space Administration.
Schor, C.M., I. Wood, and J. Ogawa
1984 Binocular sensory fusion is limited by spatial resolution. *Vision Research* 24:661-665.
Sethi, B.
1986 Heterophoria: A vergence adaptive position. *Ophthalmic Physiological Optics* 6(20):151-156.
Trevarthen, C.
1968 Two mechanisms of vision in primates. *Psychologische Forschung* 31:299-337.
Ullman, S.
1996 *High-Level Vision: Object Recognition and Visual Cognition.* Cambridge, MA: MIT Press.
U.S. Department of the Army
1993 Mission Need Statement for the Enhanced Integrated Soldier System - Dismounted (MNS). September 8, 1993 Memorandum for Assistant Secretary of the Army (Research, Development and Acquisition). Documentation No. DAMO-FDZ.
Weintraub, D.J., and M. Ensing
1992 *Human Factors Issues in Head-up Display Design: The Book of HUD.SOAR CSERIAC State of the Art.* Report number 20 92-2. Wright-Patterson Air Force Base, OH: Armstrong Medical Research Laboratory.
Wickens, C.D., and A.D. Andre
1990 Proximity, compatibility, and information display: Effects of color, space, and objectness of information integration. *Human Factors* 32:61-77.
Wickens, C.D., and J. Long
1994 Conformal symbology, attention shifts, and the head-up display. *Proceedings of the Human Factors Society 35th Annual Meeting* 38:6-10.
Wickens, C.D., R. Martin-Emerson, and I. Larish
1993 Attentional tunneling and the head-up display. *Proceedings of the 7th International Symposium on Aviation Psychology.* Department of Aviation, Ohio State University, Columbus.

CHAPTER 5

Bronkhorst, A.W., and R. Plomp

1988 The effect of head-induced interaural time and level ifferences on speech intelligibility in noise. *Journal of the Acoustical Society of America* 83:1508-1516.

1992 Effect of multiple speechlike maskers on binaural speech recognition in normal and impaired hearing. *Journal of the Acoustical Society of America* 92:3132-3139.

Deathridge, B.H.

1972 Auditory and other sensory forms of information presentation. In H.P. Van Cott and R.G. Kinkade, eds., *Human Engineering Guide to Equipment Design*. Washington DC: Government Printing Office.

Ericson, M.A., and R.L. McKinley

1996 The intelligibility of multiple talkers separated spatially in noise. In R.H. Gilkey and T.R. Anderson, eds., *Binaural and Spatial Hearing in Real and Virtual Environments*. Hillsdale, NJ: Erlbaum.

House, A.S., C.E. Williams, M.H.L. Hecker, and K.D. Kryter

1965 Articulation-testing methods: Consonantal differentiation with a closed-response set. *Journal of the Acoustical Society of America* 37:158-166.

Kistler, D.J., and F.L. Wightman

1992 A model of head related transfer functions based on principal components analysis and minimum-phase reconstruction. *Journal of the Acoustical Society of America* 91:1637-1647.

Patterson, R.D.

1982 *Guidelines for Auditory Warning Signals on Civil Aircraft*. CAA Paper 82017. London: Civil Aviation Authority.

Patterson, R.D., and R. Milroy

1980 Auditory Warnings on Civil Aircraft: The Learning and Retention of Warnings. MRC Applied Psychology Unit, Civil Aviation Authority Contract 7D/S/0142.

Pisoni, D.B.

1982 Perception of speech: The human listener as a cognitive interface. *Speech Technology* 1:10-23.

Simpson, C.A., M.E. McCauley, E.F. Roland, J.C. Ruth, and B.H. Williges

1986 Speech controls and displays. In G. Salvendy, ed., *Handbook of Human Factors*. New York: Wiley and Sons.

Sorkin, R.D.

1986 Design of auditory and tactile displays. In G. Salvendy, ed., *Handbook of Human Factors*. New York: Wiley and Sons.

Voiers, D.

1977 Diagnostic evaluation of speech intelligibility. In M.E. Hawley, ed., *Speech Intelligibility and Speaker Recognition*. Stroudsburg, PA.: Dowden, Hutchinson, and Ross.

Wightman, F.L., and D.J. Kistler

1992 The dominant role of interaural time differences in sound localization. *Journal of the Acoustical Society of America* 91:1648-1661.

CHAPTER 6

Allport, A.

1993 Attention and control: Have we been asking the wrong questions? A critical review of twenty five years. In D.E. Meyer and S. Kornblum, eds., *Attention and Performance XIV*. Cambridge, MA: MIT Press.

Beatty, J.
1982 Task-evoked pupillary responses, processing load, and the structure of processing resources. *Psychological Bulletin* 91:276-292.
Boff, K.R., and J.E. Lincoln, eds.
1988 *Engineering Data Compendium.* Wright-Patterson Air Force Base, OH: Armstrong Medical Research Laboratory.
Broadbent, D.E.
1958 *Perception and Communications.* New York: Pergamon Press.
Byers, J.C., A.C. Bittner, Jr., and S.G. Hill
1989 Traditional and raw task load index (TLX) correlations: Are paired comparisons necessary? In A. Mital, ed., *Advances in Industrial Ergonomics and Safety* (Vol. 1). London: Taylor and Francis.
Card, S.K., T.P. Moran, and A. Newell
1983 *The Psychology of Human-Computer Interaction.* Hillsdale, NJ: Lawrence Erlbaum Associates.
Carrier, M.A., and H. Pashler
1995 The attentional limits in memory retrieval. *Journal of Experimental Psychology: Learning, Memory, and Cognition* 21(5):1339-1348.
Chignell, M.H., and P.A. Hancock
1985 Knowledge-based load leveling and task allocation in human-machine systems. *Proceedings of the Annual Conference on Manual Control* 21:9.1-9.11.
1989 The operator as a purposive system: A new approach to human factors. *Proceedings of the IEEE Systems, Man and Cybernetics*, November. Boston, MA.
Damos, D.L.
1988 Individual differences in subjective estimates of workload. Pp. 231-237 in P.A. Hancock and N. Meshkati, eds., *Human Mental Workload.* Amsterdam: North-Holland.
Derrick, W.L.
1988 Dimensions of operator workload. *Human Factors* 30:95-110.
Easterbrook, J.A.
1959 The effect of emotion on cue utilization and the organization of behavior. *Psychological Review* 66:183-201.
Eberts, R.E.
1994 User Interface Design. Englewood Cliffs, NJ: Prentice Hall.
Gopher, D., and E. Donchin, E.
1986 Workload: An examination of the concept. Pp 41/1-41/49 in K. Boff, L. Kaufman, and J.P. Thomas, eds., *Handbook of Perception and Human Performance.* New York: Wiley.
Hancock, P.A.
1984 Environmental stressors. Pp. 103-142 in J.S. Warm, ed., *Sustained Attention in Human Performance.* New York: Wiley.
1986a The effect of skill on performance under an environmental stressor. *Aviation, Space, and Environmental Medicine* 57:59-64.
1986b Sustained attention under thermal stress. *Psychological Bulletin* 99:263-281.
1986c Stress and adaptability. Pp. 243-251 in G.R.J. Hockey, A.W.K. Gaillard, and M.G.H. Coles, eds., *Energetics and Human Information Processing.* Dordrecht, the Netherlands: Martinus Nijhoff.
1987 Arousal theory, stress and performance: Problems of incorporating energetic aspects of behavior into human-machine systems function. Pp. 170-179 in L.S. Mark, J.S. Warm, and R.L. Iluston, eds., *Ergonomics and Human Factors: Recent Research.* New York: Springer-Verlag.
1987 Stress and human factors in flight crew operations. *Proceedings of the Annual International Aircraft Cabin Safety Symposium* 4:17-25.

1989 The effect of performance failure and task demand on the perception of mental workload. *Applied Ergonomics* 20:197-205.

1996a On convergent technological evolution. *Ergonomics in Design* 4:(1)22-29.

1996b On the future of work. *Ergonomics in Design*, submitted April 1996.

Hancock, P.A., and M.H. Chignell

1987 Adaptive control in human-machine systems. Pp. 305-345 in P.A. Hancock, ed., *Human Factors Psychology*. Amsterdam: North-Holland.

1988 Mental workload dynamics in adaptive interface design. *IEEE Transactions on Systems, Man, and Cybernetics* 18:647-658.

Hancock, P.A., and G.R. Dirkin

1983 Stressor induced attentional narrowing: Implications for design and operation of person-machine systems. *Proceedings of the Human Factors Association of Canada* 16:19-21.

Hancock, P.A., J.A. Duley, and S.F. Scallen

1994 The Control of Adaptive Function Allocation. Report HFRL, Nav-6, September.

Hancock, P.A., and M. Vercruyssen

1988 Limits of behavioral efficiency for workers in heat stress. *International Journal of Industrial Ergonomics* 3:149-158.

Hancock, P.A., and J.S. Warm

1989 A dynamic model of stress and sustained attention. *Human Factors* 31:519-537.

Hart, S.G.

1986 Theory and measurement of human workload. Pp. 396-456 in J. Zeidner, ed., *Human Productivity Enhancement: Training and Human Factors in Systems Design* (Vol. 1). New York: Praeger.

Hart, S.G., and L. Staveland

1988 Development of the NASA task load index (TLX): Results of empirical and theoretical research. Pp. 139-183 in P.A. Hancock and N. Meshkati, eds., *Human Mental Workload*. Amsterdam: North Holland.

Hendy, K.C., K.M. Hamilton, and L.N. Landry

1993 Measuring subjective workload: When is one scale better than many? *Human Factors* 35:579-601.

Hockey, G.R.J., and P. Hamilton

1983 The cognitive patterning of stress states. In G.R.J. Hockey, ed., *Stress and Fatigue in Human Performance*. New York: Wiley.

Hoffman, J.E., M.R. Houck, F.W. MacMillan, R.F. Simons, and J.C. Oatman

1985 Event-related potentials elicited by automatic targets: A dual-task analysis. *Journal of Experimental Psychology: Human Perception and Performance* 11:50-61.

Huey, B.M., and C.D. Wickens

1993 *Workload Transitions: Implications for Individual and Team Performance*. Committee on Human Factors, Commission on Behavioral and Social Sciences and Education, National Research Council. Washington, DC: National Academy Press.

Humphrey, D.G., and A.F. Kramer

1994 Toward a psychophysiological assessment of dynamic changes in mental workload. *Human Factors* 36:3-26.

Israel, J.B., C.D. Wickens, G.L. Chesney, and E. Donchin

1980 P300 and tracking difficulty: Evidence for multiple resources in dual-task performance. *Psychophysiology* 17:259-273.

Just, M.A., and P.A. Carpenter

1992 A capacity theory of comprehension: Individual differences in working memory. *Psychological Review* 99(1):122-149.

1993 The intensity dimension of thought: Pupillometric indices of sentence processing. *Canadian Journal of Experimental Psychology* 47:310-339.

Just, M.A., P.A. Carpenter, and D.D. Hemphill

in press Constraints on processing capacity: Architectural or implementational? In D. Steir and T. Mitchell, eds., *Mind Matters: A Tribute to Allen Newell.*

Kahneman, D.

1973 *Attention and Effort.* Englewood Cliffs, NJ: Prentice-Hall.

Kieras, D.E., and D.E. Meyer

1995 An Overview of the EPIC Architecture for Cognition and Performance with Application to Human-Computer Interaction. University of Michigan, EPIC Report No. 5.

Kosslyn, S.M.

1994 *Elements of Graph Design.* New York: Freeman and Co.

Kramer, A.

1991 Physiological metrics of mental workload: A review of recent progress. In D. Damos, ed., *Multiple-Task Performance.* London: Taylor and Francis.

Lee, J., and N. Moray

1992 Trust, control strategies, and allocation of function in human machine systems. *Ergonomics* 35:12/43-12/70.

Makeig, S., and T.P. Jung

in press Changes in alertness are a principal component of variance in the EEG spectrum. *Neuroreport.*

Moray, N., B. Turksen, P. Aidie, D. Drascic, P. Eisen, E. Kruschelnicky, L. Money, H. Schonert, and C. Thornton

1986 Progress in mental workload measurement. Pp. 1121-1127 in *Proceedings of the Human Factors Society 30th Annual Meeting.*

Mulder, G.

1980 The Heart of Mental Effort: Studies in the Cardio Vascular Psychophysiology of Mental Work. State University of Groninger, The Netherlands.

Muscio, B.

1921 Is a fatigue test possible? *British Journal of Psychology* 12:31-46.

Navon, D.

1984 Resources—A theoretical soapstone? *Psychological Review* 91:216-234.

North, R.A., S.P. Stackhouse, and K. Grauffunder

1979 *Performance, Physiological, and Oculomotor Evaluation of VTOL Landing Displays.* Report No. 3171. Langley, VA.: NASA Langley Research Center.

Nygren, E.

1991 Psychometric properties of subjective workload measurement techniques: Implications for their use in the assessment of perceived mental workload. *Human Factors* 33:17-33.

O'Donnell, R.D., and F.T. Eggemeier

1986 Workload assessment methodology. Pp. 42:1-49 in K. Boff, L. Kaufman, and J.P. Thomas, eds., *Handbook of Perception and Human Performance.* New York: Wiley.

Pashler, H.

1989 Dissociations and dependencies between speed and accuracy: Evidence for a two-component theory of divided attention in simple tasks. *Cognitive Psychology* 21:469-514.

1991 Shifting visual attention and selecting motor responses: Distinct attentional mechanisms. *Journal of Experimental Psychology: Human Perception and Performance* 17:1023-1040.

1994 Dual-task interference in simple tasks: Data and theory. *Psychological Bulletin* 116:220-244.

Pashler, H., M. Carrier, and J.E. Hoffman
1993 Saccadic eye movements and dual-task interference. *Quarterly Journal of Experimental Psychology* 46:51-82.

Porges, S.
1985 Respiratory sinus arrhythmia; An index of vagal tone. In Mulder and Van Doornen, eds., *Psychophysiology of Cardiovascular Control*. New York: Plenum Press.

Riley, V.
1994 Human Use of Automation. Unpublished doctoral dissertation, University of Minnesota, Minneapolis.

Ruthruff, E., J. Miller, and T. Lachman
1995 Does mental rotation require central mechanisms? *Journal of Experimental Psychology: Human Perception and Performance* 21(3):552-570.

Sampson, J.B.
1993 Cognitive performance of individuals using a head-mounted display while walking. Pp. 338-341 in *Proceedings of the Human Factors Society 37th Annual Meeting*.

Sanders, A.F.
1983 Toward a model of stress and human performance. *Acta Psychologica* 53:61-97.

Schneider, W.
1985 Training high-performance skills: Fallacies and guidelines. *Human Factors* 27:285-300.

Schneider, W., and R.M. Shriffin
1977 Controlled and automatic human information processing (I): Detection, search, and attention. *Psychological Review* 84:1-66.

Shah, P., and P.A. Carpenter
1995 Conceptual limitations in comprehending line graphs. *Journal of Experimental Psychology: General* 124:43-61.

Shah, P., and A. Miyake
in press The separability of working memory resources for spatial thinking and language processing: An individual differences approach. *Journal of Experimental Psychology: Learning, Memory, and Cognition*.

Sirevaag, E.J., A.F. Kramer, C.D. Wickens, M. Reisweber, D.L. Strayer, and J.F. Grenell
1993 Assessment of pilot performance and mental workload in rotary wing aircraft. *Ergonomics* 36:1121-1140.

Stern, J.A., and J.J. Skelly
1984 The eye blink and workload considerations. Pp. 942-944 in *Proceedings of the Human Factors Society 28th Annual Meeting*.

St. George, M., J.E. Hoffman, and S.M. Mannes
1995 Individual Differences in Inference Making: An ERP Analysis. Paper presented at the 36th annual meeting of the Psychonomic Society, Los Angeles CA, November.

Travis, D.
1991 *Effective Color Displays*. New York: Academic Press.

Treisman, A., and G. Gelade
1980 A feature-integration theory of attention. *Cognitive Psychology* 12:97-136.

Vercruyssen, M., P.A. Hancock, and T. Mihaly
1988 Behavioral toxicology: Effects of breathing 4% carbon dioxide on cognition. *Proceedings of the International Ergonomics Association*, August. Sydney, Australia.

Wickens, C.D.
1980 The structure of attentional resources. In R.S. Nickerson, ed., *Attention and Performance, VIII*. Hillsdale, NJ: Erlbaum.
1984 Processing resources in attention. In R. Parasuraman and R. Davies, eds., *Varieties of Attention*. New York: Academic Press

1992 *Engineering Psychology and Human Performance* (2nd ed.). Scranton, PA: Harper Collins.

Wickens, C.D., and C.M. Carswell
1995 *Human Factors* 37:473-494.

Wickens, C.D., R. Martin-Emerson, and I. Larish
1993 Attentional tunneling and the head-up display. *Proceedings of the 7th International Symposium on Aviation Psychology.* Columbus, OH: Dept. of Aviation, Ohio State University.

Wolfe, J.M.
1994 Guided Search 2.0: A revised model of visual search. *Psychonomic Bulletin and Review* 1(2):202-238.

Yeh, Y.Y., and C.D. Wickens
1988 Dissociation of performance and subjective measures of workload. *Human Factors* 30:111-120.

Yerkes, R.M., and J.D. Dodson
1908 The relation of strength of stimulus to rapidity of habit formation. *Journal of Comparative and Neurological Psychology* 18:459-482.

CHAPTER 7

Vreuls, D., and R.W. Obermayer
1985 Human-system performance measurement in training simulators. *Human Factors* (Special Issue: Training) 27(3):241-250.

CHAPTER 8

CuQlock-Knopp, V.G., W. Torgerson, D.E. Sipes, E. Bender, and J.O. Merritt
1994 Comparison of Monocular, Biocular, and Binocular Night Vision Goggles for Traversing Off-Road Terrain on Foot. ARL-TR-747, March. Aberdeen Proving Ground, MD: U.S. Army Research Laboratory.

APPENDIXES

APPENDIX A
Control Ergonomics for the Helmet-Mounted Display

In the course of its work, the panel did not address directly the question of control design for the helmet-mounted display. Because the functions to be controlled are not yet fully defined, it was not an appropriate question for the panel. However, our review of battlefield conditions, weapons deployment, and other matters related to the use of the helmet-mounted display generated some concerns on the part of panel members about the ergonomics of the controls to be used in the Land Warrior System. These concerns are covered briefly in this appendix.

BACKGROUND

In the battlefield setting of the past, the principal control used by the infantry soldier was the trigger on a gun. Even in relatively recent years, although weapons have become progressively more complicated, the controls used in the operation of these advanced weapons have been kept as simple as possible. With the advent of systems such as Land Warrior and its helmet-mounted display subsystem, however, vastly different levels of complexity have been introduced. As a result, the problem of control design for infantry weapon systems has become much more difficult than it once was.

Some of these new difficulties may be overcome by a reasonable adherence to the basic human factors principles that are available in handbook form (see e.g., Military Standard 1472-D). An expanded assembly of control design guidelines may also be needed to address the combination of sophisticated systems with the environmental stresses of combat on the ground by individual dismounted soldiers.

DESIGN OBJECTIVES

Some of the design objectives will derive from decisions on the military doctrine that will stipulate the manner of use of the helmet-mounted display. Although such doctrine is not yet available, there are both analytical and empirical issues that can be raised that can give some orientation to the design effort.

As a point of departure and focus, the prime concern probably should be the digital computer component of the Land Warrior System. Even with this focus, there is some ambiguity about function. For example, it appears that one computer function will be to process signals—in real time—from higher command sources. One specific class of messages might cover the disposition of hostile forces in the immediate area. However, the precise form of such information is not yet known, and many possible message categories have yet to be stipulated. Nevertheless, it is clear that controls will be needed to regulate both the computer data processing and the display functions. It also seems clear that such controls are unlikely to be in the same family of devices that are used to regulate the channel, the brightness, and the contrast of video signals on the typical home television screen.

At least some of the adjustable attributes of the display should probably be preset and then maintained by the computer itself, with the additional provision for emergency manual override capabilities. This option is warranted on the grounds of minimizing task load on the soldier.

A related control design problem emerges from the prospect that the Land Warrior System can provide a choice of optical enhancements to the soldier. The key design question is not just the mode of actuation of one form of optical enhancement and the deactivation of another. It is that, given the presence of computer capability, can it be used to facilitate the soldier's adjustment to the changeover?

Another array of functions that are likely to be implemented in the Land Warrior System are location determination, route specification and ranging for direct and indirect fire. These functions stem from the capabilities provided by the global positioning system. Fortunately, the control requirement for these functions is relatively simple—activation/deactivation for location and route display. However, if a non-preplanned route is sought, something more akin to the query structure built into advanced personal computers might be needed. In other words, the soldier, as user, will need to be able to tell the computer both destination parameters and route information, such as a request for a route that does not involve visual exposure to an opposing force. Point-and-click procedures using a body-mounted trackball device may not be optimal for battlefield use because of dexterity problems. Likewise, a voice-activated control has drawbacks in conditions of high background noise levels and in situations in which stealth is desired.

STIPULATIONS AND CRITERIA

Questions that are based on combat operations and that are subject to analytic resolutions include the following:

1. What functions of the system require control?
2. Should the exercise of control (of a particular function) be allocated to the individual soldier or to a computer?
3. If allocated to a computer, should an override option be provided?

At some stage in the development of the Land Warrior System, attention should be focused on circumstances in which the dismounted soldier will face chemical, biological, and radiological weapons. Likewise, such features as automatic land mine detection should be considered.

A second set of questions will require research to resolve. Because of the complexity of the Land Warrior System, not only are there many single variables that need assessment by research but also many trade-off functions and interactions will require systematic study in a field setting. In that regard, there are basic questions about how a given control is to be put to use as well as additional crucial questions about how one mode of use relates to the other modes.

It seems unlikely that there is a single location (wrist, helmet, chest, belt, weapon stock, etc.) where the full complement of controls can be located without penalty. It seems equally unlikely that any one mode (keyboard, trackball, voice, etc.) will provide the ideal means of control. However, trying various arrangements in field or field-like conditions is a relatively straightforward test project that could lead directly to a minimally disruptive array of control locations.

Such an effort would be congruent with the thrust of the overall Land Warrior/helmet-mounted display program—which is to give the dismounted soldier as much of a tactical advantage as possible while not adding to his problems. This general goal also leads to some reasonable specifications for the designers of the controls.

First, the controls should be kept as simple (and rugged) as possible. They should also be protected from inadvertent activation—by the soldier or by obstructions in the environment—but at the same time should be easily and quickly accessible. Whenever possible, there should be strong cues to the function over which the control presides. Such cues include location in sets, proximity to the device being controlled, and some easy abstraction such as a shape cue or a color coding that is not ambiguous (i.e., red = stop).

These and other standard ergonomic stipulations allow great discretion on the part of the designers—so they should not be a source of constraints or inhibitions. However, no matter how elaborate the design guidance, there will always remain some uncertainties related to the specific requirements generated by the overall system concept and its interactions with the populations of human users and the contexts within which use will take place. Uncertainties that are system-specific are the warrant for sustaining a strong research, test, and evaluation capability. Many trade-off studies will be needed before the Land Warrior equipment is ready to issue to the troops. The framework for such research should be a strong user orientation. Resolving the uncertainties of control design should be driven by the soldier's sense of what is needed, useful, and preferable when going into combat.

APPENDIX

B

Physical Ergonomics of the Infantry Helmet

The protective helmet currently in use by infantry soldiers in the U.S. Army was developed in the early 1970s and was last tested for its ergonomic suitability in 1976 (McManus et al., 1976). In the intervening 20 years, crucial advances have taken place in the technologies of ballistic protection and in the design of displays for military use. In particular, display improvements have been made in the capability to enhance vision under low-light conditions by the use of photomultiplication technology. The size and weight of such devices have been greatly reduced—making it possible to provide a highly portable night-vision apparatus for use by dismounted soldiers. A secondary benefit of such equipment is that it can be made sensitive to infrared frequencies. This means that objects that radiate in the infrared range can be discerned by the appropriately equipped soldier in low ambient light conditions. Furthermore, high-intensity infrared sources now can be used for communication purposes or to illuminate targets without revealing intent or position to hostile forces that are not so equipped. In short, technical means now are available to conduct many operations that once were restricted to daylight execution.

In addition to image enhancement, electronic technologies have improved the capability to convey both real scenic images (e.g., television pictures), symbol sets (e.g., maps and instrument readings), and narrative text on a miniature opaque screen.

The combination of these technical advances has made it possible to contrive small, multipurpose, multimode displays that can be mounted on a headband or a helmet. Applications of this capability have been undertaken for crew members in both fixed and rotary wing aircraft and in some surface vehicles. The step toward providing similar benefit for the dismounted soldier was a reasonable and logical extension of the applications already shown to be successful. However, the research, development, test, and evaluation work that has been done on air crew (Rash et al., 1987) and tanker helmet-mounted displays (Nelson, 1994) is not particularly relevant to the problems associated with the use of such equipment by dismounted soldiers. The weight of the equipment can be externally supported in a vehicle, and personal mobility is not usually a requirement for such vehicle-borne personnel. Consequently, the design of a helmet-mounted display for the dismounted soldier presents a set of new problems for engineers and program managers.

ERGONOMICS DESIGN CRITERIA

The infantry soldier differs from the other fighting personnel because he must physically carry all of his equipment. That means that weight is a design criterion for the helmet-mounted display (Kibler, 1975). Although serious effort has been given to minimizing the weight of the helmet-mounted display for the Land Warrior System, it still represents some increment over what the soldier must carry now—and that constitutes an ergonomic drawback. From this perspective, an increment of more than approximately 1 kilogram on the helmet and possibly 2 kilograms in a back or side pack would generate measurable impairment to the endurance of the soldier (Woodson and Conover, 1964).

In addition to sheer weight, other key criteria are as follows:

Weight distribution. The central issue in weight distribution is the possibility of a shift in the center of gravity—both with respect to the head alone and to the body as a whole. Any shift in the weight distribution on the head away from its normal center of gravity can increase the likelihood of whiplash-type injuries to the neck (Jones et al., 1972). Some muscle strain will accompany any displacement—and the strain will be proportional to the lever force on the neck (i.e., the force vector composed of the weight and the angle of displacement). Changes in overall bodily center of gravity impair balance during movement. The compounded problem in this area comes from the fact that the body has reflexes and highly overlearned muscular compensation responses when moving at different speeds. Thus, whereas the ideal location of any weight on the head might be centered on the vertical line connecting the body's center of gravity and the head's center of gravity, the situation is complicated by the postural changes as one shifts from an erect, standing position to a slow walk, to a fast walk, to a jog, and to a run. In combat, when the soldier may be in a partial crouch while moving as quickly as possible, the problem of maintaining balance will require extensive muscular accommodations with the cost of added stress and fatigue.

Protection. If a new helmet is developed for the dismounted soldier, all standard protection tests will need to be conducted (Perry, 1994). Ballistic protection is a primary consideration. The capabilities of the current helmet are good for fragments and general debris, but it will not deflect a direct hit from a standard military rifle round (Corona et al., 1974). There are several facets to the protection criterion that suggest substantial inquiries by empirical means. One is the evaluation of new materials that might be superior to that currently used, Kevlar. For example, some preliminary tests of Aramid as a Kevlar substitute have already been conducted (North Atlantic Treaty Organization, 1995). In any case, the 20-year interval from the adoption of the current helmet leads to the thought that some advances in materials or concepts are likely. Such advances have taken place in the design of protective headgear for various sports (Hurt and Thom, 1993). The differences in the headgear worn by bicyclists and that worn by football players is instructive. For example, football helmets are designed for sustained use, whereas cyclists' helmets are designed to serve in only a single incident (Vetter et al., 1987).

Given the long interval since the last redesign of the infantry helmet, it may be useful to consider what standards of ballistic protection are valid in present-day war-fighting environments (see, e.g., Edwards and Kash, 1995).

Freedom of head movement. In combat settings, the individual soldier must be able to visually scan the entire scenic surround for threats and targets. To do so efficiently and with minimum exposure, the head must be free so that head-neck movements are totally unimpaired. Even if the shape of the helmet itself does not restrict such movement (U.S. Army Human Engineering Laboratory, 1973; Scheetz et al., 1973), care must be exercised with respect to any attachments to the helmet. Of particular concern should be the use of wires or cables that provide electronic connections from the helmet-mounted display with other equipment such as remote gun sights and navigation gear such as the global position system and computer equipment. Wires or cables can become snagged on other items of carried gear as well as on external objects such as vegetation. Various distributions of the cabling connections should be tried under field conditions to determine what arrangement is the least likely to give the soldier problems in head movement.

Helmet position stability. In electro-optical systems in which the eye is positioned a few centimeters from a small display, there is little or no tolerance for any shift in the relative location of the display with respect to the eye. In terms of physical ergonomics, this means that the helmet cannot be free to move on the wearer's head—or if some movement is allowed, resettling of the helmet must be very easy and quickly accomplishable by the wearer. A potentially detrimental trade-off is presented by the prospect that restrictions on helmet-to-head movement will mean the use of some kind of harness that will add weight and possibly be quite uncomfortable to the wearer. Very little research has been done on headgear restraints. The modest data that are available indicate that an adequate design solution will be difficult to achieve when mobility factors are also taken into account.

Micro-climate control. Various options exist with respect to the arrangement of support for the helmet or helmet liner (Fonseca, 1974). Webbing, foam padding, liquid-filled pads, and other materials provide some reasonable alternatives. However, those that might fulfill the need for weight distribution and positional stability are likely to be deficient with respect to air circulation inside the helmet. It appears likely that the optimally stabilized helmet will add to the heat load of the wearer. This may not be too debilitating except that some wearers will produce quantities of sweat that can impair vision. A possible counteraction would be to equip the helmet with some form of cooling capability; however, the trade-off is added weight.

User acceptance. User acceptance of a newly designed helmet depends in part on the actual quality of the fit (Robinette, 1993) and freedom from discomfort (Mozo et al., 1995). However, research on user acceptance of innovations strongly suggests that users conduct a form of subjective cost-benefit analysis in forming their reactions. Another important variable is the user's sense of participation or influence on the configuration of the innovation. And group effects such as the expressed attitudes of fellow workers can be strong (Coch and French, 1948). Engaging prospective users early and often in the development process is one way to promote user acceptance of the final product. This procedure is even more effective if participation is shared across the workforce—or if a large, representative sample of users is engaged in the early evaluation stages of innovation development. The crucial factor is whether or not the benefits objectively outweigh the costs to the user. For the helmet-mounted display, penalties in the form of added weight and physical stress are inevitable. For acceptance to be achieved, it must be clearly evident to the individual soldier that the device is highly beneficial—not just with respect to the likelihood of mission accomplishment—but also with respect to survivability. If the helmet-mounted display is perceived as providing a better chance of completing an engagement without being killed or wounded, the soldier will tolerate some added stress.

RESEARCH NEEDS

The lack of system-specific research on helmets and helmet attachments for dismounted soldiers means that the configuration of the Land Warrior System will probably be based on the judgements of the design engineers and extrapolations from adjacent research areas. The lack of direct comparability between the research on helmets and helmet-mounted displays for vehicular crew members and the conditions that are experienced by the dismounted soldier provides a strong argument for a comprehensive program of laboratory research augmented by small-scale simulated engagement tests.

The laboratory tests should be mainly directed to the helmet per se and the liner and support cushioning materials. Small-scale operational simulation means the use of a design prototype or a succession of prototypes to outfit a small unit, such as a fire team, which is then put through some exercises that are roughly comparable to an engagement situation. The purpose of tests of this kind is to identify particular problems—not to generate solutions. For example, if the helmet attachments project to the front of the soldier's face, will he adjust his movements to avoid impacts between the device and the ground? Also, the researchers should acquire a fair notion of what features of the helmet and its attachments give rise to negative feelings or complaints on the part of the wearers. Ultimately, such-small scale field tests should result in a user-centered configuration.

DESIGN AND DEVELOPMENT GUIDELINES

There is very little theory to guide the program of research on the physical ergonomics of the helmet and its attachments. However, human factors design principles can be interpreted to fit the specific ergonomic issues posed by the proposed system:

The helmet and its attachments should be as light weight as possible while fulfilling the objective of providing ballistic protection.

The attachments, including cables and wires, should extend as short a distance as possible from the head and body of the soldier.

Any attachments to the helmet should be easily removable by the soldier in a field environment.

The helmet should be cushioned so that the contours automatically adjust to the conformation of the wearer's head.

The positioning of the helmet should be as stable as possible while allowing free head movement and either air circulation around the head, miniaturized climate control, or both.

CONCLUSION

The base of engineering design and test data for the development of a helmet-mounted (or hand-carried) display subsystem to be used by dismounted soldiers is not sufficient for the purpose. A program of system-specific laboratory research and prototype testing could significantly enhance the likelihood that the Land Warrior ensemble will meet all military criteria.

REFERENCES

Coch, L., and J.P.R. French
1948 Overcoming resistance to change. *Human Relations* 1:512-532.

Corona, B.M., P.H. Ellis, R.D. Jones, R.B. Randall, and H.A. Scheetz
1974 *Human Factors Evaluation of Two Proposed Army Infantry/Marine Fragmentation Protective Systems*. TM24-74. Aberdeen Proving Ground, MD: U.S. Army Human Engineering Research Laboratory.

Edwards, A.G., IV, and H.M. Kash, III
1995 *Flechettes Against Body Armor: Effects of Varying Area of Coverage on Incapacitation and Survivability*. ARL-TN-54. Aberdeen Proving Ground, MD: U.S. Army Human Research Laboratory.

Fonseca, G.F.
1974 *Heat Transfer Properties of Military Protective Headgear.* TR 74-29-CE. Natick, MA: U.S. Army Natick Laboratories.
Hurt, H.H., Jr., and D.R. Thom
1993 Accident performance of motorcycle and bicycle safety helmets. In E.F. Hoerner, ed., *Head and Neck Injuries in Sports.* ASTM STP 1229. Philadelphia: American Society for Testing and Materials.
Jones, R.D., B.M. Corona, P.H. Ellis, R.B. Randall, and H.A. Scheetz
1972 *Perception of Symmetrically Distributed Weight on the Head.* TN 4-72. Aberdeen Proving Ground, MD: U.S. Army Human Engineering Research Laboratory.
Kibler, R.J.
1975 *A System for Determining Helmet Moments of Inertia.* TN 6-75. Aberdeen Proving Ground, MD: U.S. Army Human Engineering Research Laboratory.
McManus, L.R., P.E. Durand, and W.D. Claus, Jr.
1976 *Development of a New Infantry Helmet.* TR 76-30-CEMEL. Natick, MA: U.S. Army Natick Research Development and Engineering Center.
Mozo, B.T., B.A. Murphy, and J.E. Ribera
1995 *User Acceptability and Comfort of the Communication Earplug (CEP) When in Use in the UH-1 Helicopter.* Report No. 95-17. Fort Rucker, AL: U.S. Army Aeromedical Research Laboratory.
Nelson, S.A.
1994 CVC helmet-mounted display—next generation high resolution head-mounted display. In R.J. Lewandowski, W. Stephen, and L.A. Haworth, eds., *Helmet- and Head-Mounted Displays and Symbology Design Requirements.* Proceedings of the International Society of Optical Engineers 2218. Bellingham, WA: ISOE.
North Atlantic Treaty Organization
1995 *Dismounted Personnel Target.* Doc. AC/225-D/1328. Brussels, Belgium: North Atlantic Council (NATO).
Perry, C.E.
1994 *Vertical Impact Testing of Two Helmet-Mounted Night Vision Systems.* Wright-Patterson Air Force Base, OH: Armstrong Medical Research Laboratory.
Rash, C.E., J.S. Martin, D.W. Gower, Jr., J.R. Licina, and J.V. Barson
1987 *Evaluation of the U.S. Army Fitting Program for the Integrated Helmet Unit of the Integrated Helmet and Display Sighting System.* Fort Rucker, AL: U.S. Army Aeromedical Research Laboratory.
Robinette, K.M.
1993 Fit testing as a helmet development tool. Pp. 69-73 in *Proceedings of Human Factors Society 37th Annual Meeting*, Santa Monica, CA.
Scheetz, H.A., B.M. Corona, P.H. Ellis, R.D. Jones, and R.B. Randall
1973 *Method for Human Factors Evaluation of Ballistic Protective Helmets.* Aberdeen Proving Ground, MD: U.S. Army Human Engineering Research Laboratory.
U.S. Army Human Engineering Laboratory
1973 *Summary of Infantry Helmet Edge-Cut Criteria.* PASGT Ref. HLR-7 Helmet. Aberdeen Proving Ground, MD: U.S. Army Human Engineering Research Laboratory.
Vetter, L., R. Vanderby, and L.J. Broutman
1987 Influence of materials on performance of a football helmet. *Polymer Engineering Science* 27(15):1113-1120.
Woodson, W.E., and D.W. Conover, eds.
1964 *Human Engineering Guide to Equipment Design.* Berkeley, CA: University of California Press.

APPENDIX C

Biographical Sketches

WILLIAM O. BLACKWOOD is vice president with Hay Management Consultants and has over 25 years of human resource planning and management experience. He has directed and managed organizational and managerial analyses, planned and successfully implemented organizational change projects, designed and established total quality management (TQM) programs, and established successful strategic planning processes. He is currently working on the implementation and assessment of new work processes associated with strategic planning initiatives or TQM implementation efforts. He was the principal author and architect of the U.S. Army program called MANPRINT, which integrates manpower, personnel, training, system safety, health hazard assessment, and human factors into the system acquisition process in order to improve human performance, manpower utilization, and organizational effectiveness. Prior to joining Hay, he was the manager of the Human Systems Department of PRC Inc. He has Ph.D. and M.Ed. degrees in educational administration from the University of Florida, and a B.S. in psychology and education from Norwich University. He is a trustee for Norwich University, a fellow with the Inter-University Seminar on Armed Forces and Society, and a member of the Human Factors and Ergonomics Society.

TIMOTHY R. ANDERSON is on the staff of the Air Force Aerospace Medical Research Laboratory at Wright Patterson Air Force Base, Ohio, as well as adjunct professor at the Air Force Institute of Technology and Wright State University. His research interests include speech recognition, speaker recognition and verification, auditory modeling for speech processing, binaural hearing, binaural speech recognition, and neural networks. Anderson is the author of numerous papers and coeditor of *Binaural and Spatial Hearing in Real and Virtual Environments.* He has a B.S.E.E. from the University of Kentucky, an M.S.E.E. from Purdue University, and a Ph.D. in electrical engineering from the University of Dayton.

C. THOMAS BENNETT retired from the Army after 20 years in the Medical R&D Command working on the behavioral neuropharmacology of nerve agent antidotes and radiation protective drugs; human factors analysis of large military systems and force structures; and virtual flight displays and the graphical information required for flight control. During the last five years at Lawrence Livermore National Laboratory, he has been involved in developing the Department of Energy draft standards for human factors design methods; designing and testing tools to identify the risk factors that human operators and organizations bring to the disassembly of nuclear weapons; writing the human factors functional requirements for the next generation airport security checkpoints; and developing a human and organizational systems analytic basis for the Concept of Operations document that will be used to run the uranium enrichment plant being developed by the United States Enrichment Corporation. He has an undergraduate degree from the University of Michigan and a Ph.D. in psychology from Michigan State University.

JOHN R. CORSON is currently a consultant to the U.S. Army in the areas of training and test and evaluation. His experience is in the fields of acquisition and management, training development and operational test and evaluation, as well as operations research, training analysis and design and doctrine analysis. Corson previously was chief executive officer at Integrated Visual Learning, vice president of Mandex, Inc., and deputy commander, U.S. Army Operational Test and Evaluation Agency. He has a B.S. in industrial management from Drexel University and an M.B.A. in logistics management from Ohio State University.

MICA R. ENDSLEY is currently visiting associate professor at Massachusetts Institute of Technology in the Department of Aeronautics and Astronautics and associate professor of industrial engineering at Texas Tech University. She has been conducting research on situation awareness, decision making and automation in high-performance aircraft, and air traffic control and maintenance for the past 11 years. Prior to joining Texas Tech in 1990, she was an engineering specialist for the Northrop Corporation, serving as principal investigator of a research and development program focused on the areas of situation awareness, mental workload, expert systems, and interface design for the next generation of fighter cockpits. She has a Ph.D. in industrial and systems engineering from the University of Southern California, with a specialization in human factors, and B.A. and M.A. degrees from Texas Tech University and Purdue University, respectively, both in industrial engineering. She has published extensively on the topic of situation awareness and designing for automation and is a registered professional ergonomist.

PETER A. HANCOCK is associate professor at the University of Minnesota and director of the Human Factors Research Laboratory there. From 1983 to 1989 he was assistant professor in the Departments of Safety Science and Human Factors at the University of Southern California. His current research concerns operator strategies in performance under time stress and the application of safety and human factors principles to the design and implementation of advanced transportation systems. He has edited three books, *Human Factors Psychology*, *Human Mental Workload*, and *Intelligent Interfaces: Theory, Research and Design*, the latter two in association with Najmedin Meshkati and Mark Chignell. He is the coeditor of two forthcoming texts on the application of ecological principles to human factors. He has a B.Ed and M.Sc. from Loughborogh University with a focus on computer modeling of physiological systems. He received a Ph.D. in 1983 from the University of Illinois, where his research concerned the human perception of time.

JULIAN E. HOCHBERG is Centennial Professor Emeritus of psychology at Columbia University. His research interests include the perceptual integration of successive glances at objects and scenes; perception and attention; and pictorial perception and communication. He was elected to the National Academy of Sciences in 1980. He has an M.A. and Ph.D. in psychology from the University of California, Berkeley.

JAMES E. HOFFMAN is professor of psychology at the University of Delaware. His research has been primarily in the area of visual attention and search using behavioral, electrophysiological, and eye movement recording approaches. In summer 1996, he was visiting research scientist at the Night Vision Laboratory in Fort Belvoir, Virginia, where he investigated eye fixation patterns during target detection in flare-illuminated scenes. His current work involves extending this approach to search and detection in thermal imagery. He has B.A. and M.A. degrees in psychology and a Ph.D. in cognitive psychology from the University of Illinois, Champaign/Urbana.

JERRY KIDD is senior adviser for the Committee on Human Factors and its various projects. He received a Ph.D. from Northwestern University in social psychology in 1956; he then joined RAND Corporation to help on a project to simulate air defense operations. He left RAND in late 1956 to join the staff at the Laboratory of Aviation Psychology at Ohio State University. There he worked under Paul Fitts and George Briggs until 1962, when he joined the staff of AAI, Incorporated, north of Baltimore, Maryland. In 1964, he moved to the National Science Foundation as program director for special projects. He joined the faculty of the College of Library and Information Services at the University of Maryland in 1967 and retired in 1992.

RONALD V. KRUK received a B.A. and M.A. in psychology from the University of Manitoba and a Ph.D. in applied psychology from Dalhousie University. He is currently manager of collaborative research and development programs at CAE Electronics Ltd., St. Laurent, Quebec, Canada. In prior military service with the Canadian Forces he was trained and served as a pilot on both fixed-wing aircraft and helicopters. Since joining CAE Kruk has been involved in the design and development of computer image generation systems and visual displays, including helmet-mounted displays for simulation, telerobotics, and enhanced/synthetic vision systems. His current responsibilities include design and coordination of research and development activity in human factors (controls and displays), vision (human performance) and training (training system analysis, requirements definition).

ANNE MAVOR is study director for the Committee on Human Factors and its panels on tactical display systems for the individual soldier, human factors in air traffic control, and modeling human behavior and command decision making. Her previous work as a National Research Council senior staff officer includes studies of the scientific and technological challenges of virtual reality, emerging needs and opportunities for human factors research, modeling cost and performance of military enlistment, and others. For the past 25 years her work has concentrated on human factors, cognitive psychology, and information system design. Prior to joining the National Research Council she worked for the Essex Corporation, a human factors research firm, and served as a consultant to the College Board. She has an M.S. in experimental psychology from Purdue University.

Index

A

B

C

D

E

F

G

H

I

L

M

N

O

T

U

V

W